T0213537

Deep Generative Modeling

Jakub M. Tomczak

Deep Generative Modeling

 Springer

Jakub M. Tomczak
Vrije Universiteit Amsterdam
Amsterdam
Noord-Holland, The Netherlands

ISBN 978-3-030-93160-5   ISBN 978-3-030-93158-2   (eBook)
https://doi.org/10.1007/978-3-030-93158-2

This Springer imprint is published by the registered company Springer Nature Switzerland AG
The registered company address is: Gewerbestrasse 11, 6330 Cham, Switzerland

*To my beloved wife Ewelina,
my parents, and brother.*

Foreword

In the last decade, with the advance of deep learning, machine learning has made enormous progress. It has completely changed entire subfields of AI such as computer vision, speech recognition, and natural language processing. And more fields are being disrupted as we speak, including robotics, wireless communication, and the natural sciences.

Most advances have come from supervised learning, where the input (e.g., an image) and the target label (e.g., a "cat") are available for training. Deep neural networks have become uncannily good at predicting objects in visuals scenes and translating between languages. But obtaining labels to train such models is often time consuming, expensive, unethical, or simply impossible. That's why the field has come to the realization that unsupervised (or self-supervised) methods are key to make further progress.

This is no different for human learning: when human children grow up, the amount of information that is consumed to learn about the world is mostly unlabeled. How often does anyone really tell you what you see or hear in the world? We must learn the regularities of the world unsupervised, and we do this by searching for patterns and structure in the data.

And there is lots of structure to be learned! To illustrate this, imagine that we choose the three colors of each pixel of an image uniformly at random. The result will be an image that with overwhelmingly large probability will look like gibberish. The vast majority of image space is filled with images that do not look like anything we see when we open our eyes. This means that there is a huge amount of structure that can be discovered, and so there is a lot to learn for children!

Of course, kids do not just stare into the world. Instead, they constantly interact with it. When children play, they test their hypotheses about the laws of physics, sociology, and psychology. When predictions are wrong, they are surprised and presumably update their internal models to make better predictions next time. It is reasonable to assume that this interactive play of an embodied intelligence is key to at least arrive at the type of human intelligence we are used to. This type of learning has clear parallels with reinforcement learning, where machines make plans, say to play a game of chess, observe if they win or lose, and update their models of the world and strategies to act in them.

But it's difficult to make robots move around in the world to test hypotheses and actively acquire their own annotations. So, the more practical approach to learning with lots of data is unsupervised learning. This field has gained a huge amount of attention and has seen stunning progress recently. One only needs to look at the kind of images of non-existent human faces that we can now generate effortlessly to experience the uncanny sense of progress the field has made.

Unsupervised learning comes in many flavors. This book is about the kind we call probabilistic generative modeling. The goal of this subfield is to estimate a probabilistic model of the input data. Once we have such a model, we can generate new samples from it (i.e., new images of faces of people that do not exist).

A second goal is to learn abstract representations of the input. This latter field is called representation learning. The high-level representations self-organize the input into "disentangled" concepts, which could be the objects we are familiar with, such as cars and cats, and their relationships.

While disentangling has a clear intuitive meaning, it has proven to be a rather slippery concept to properly define. In the 1990s, people were thinking of statistically independent latent variables. The goal of the brain was to transform the highly dependent pixel representation into a much more efficient and less redundant representation of independent latent variables, which compresses the input and makes the brain more energy and information efficient.

Learning and compression are deeply connected concepts. Learning requires lossy compression of data because we are interested in generalization and not in storing the data. At the level of datasets, machine learning itself is about transferring a tiny fraction of the information present in a dataset into the parameters of a model and forgetting everything else.

Similarly, at the level of a single datapoint, when we process for example an input image, we are ultimately interested in the abstract high-level concepts present in that image, such as objects and their relations, and not in detailed, pixel-level information. With our internal models we can reason about these objects, manipulate them in our head and imagine possible counterfactual futures for them. Intelligence is about squeezing out the relevant predictive information from the correlated soup pixel-level information that hits our senses and representing that information in a useful manner that facilitates mental manipulation.

But the objects that we are familiar with in our everyday lives are not really all that independent. A cat that is chasing a bird is not statistically independent of it. And so, people also made attempts to define disentangling in terms of (subspaces of variables) that exhibit certain simple transformation properties when we transform the input (a.k.a. equivariant representations), or as variables that one can independently control in order to manipulate the world around us, or as causal variables that are activating certain independent mechanisms that describe the world, and so on.

The simplest way to train a model without labels is to learn a probabilistic generative model (or density) of the input data. There are a number of techniques in the field of probabilistic generative models that focus directly on maximizing the log-probability (or a bound on the log probability) of the data under the generative model. Besides VAEs and GANs, this book explains normalizing flows, autoregressive models, energy-based models, and the latest cool kid on the block: deep diffusion models.

One can also learn representations that are good for a broad range of subsequent prediction tasks without ever training a generative model. The idea is to design tasks for the representation to solve that do not require one to acquire annotations. For instance, when considering time varying data, one can simply predict the future, which is fortunately always there for you. Or one can invent more exotic tasks such as predicting whether a patch was to the right of the left of another patch, or whether a movie is playing forward or backward, or predicting a word in the middle of a sentence from the words around it. This type of unsupervised learning is often called self-supervised learning, although I should admit that also this term seems to be used in different ways by different people.

Many approaches can indeed be understood in this "auxiliary tasks" view of unsupervised learning, including some probabilistic generative models. For instance, a variational autoencoder (VAE) can be understood as predicting its own input back by first pushing the information through an information bottleneck. A GAN can be understood as predicting whether a presented input is a real image (datapoint) or a fake (self-generated) one. Noise contrastive estimation can be seen as predicting in latent space whether the embedding of an input patch was close or far in space and/or time.

This book discusses the latest advances in deep probabilistic generative models. And it does so in a very accessible way. What makes this book special is that, like the child who is building a tower of bricks to understand the laws of physics, the student who uses this book can learn about deep probabilistic generative models by playing with code. And it really helps that the author has earned his spurs by having published extensively in this field. It is a great to tool to teach this topic in the classroom.

What will the future of our field bring? It seems obvious that progress towards AGI will heavily rely on unsupervised learning. It's interesting to see that the scientific community seems to be divided into two camps: the "scaling camp" believes that we achieve AGI by scaling our current technology to ever larger models trained with more data and more compute power. Intelligence will automatically emerge from this scaling. The other camp believes we need new theory and new ideas to make further progress, such as the manipulation of discrete symbols (a.k.a. reasoning), causality, and the explicit incorporation of common-sense knowledge.

And then there is of course the increasingly important and urgent discussion of how humans will interact with these models: can they still understand what is happening under the hood or should we simply give up on interpretability? How will our lives change by models that understand us better than we do, and where humans who follow the recommendations of algorithms are more successful than those who resist? Or what information can we still trust if deepfakes become so realistic that we cannot distinguish them anymore from the real thing? Will democracy still be able function under this barrage of fake news? One thing is certain, this field is one of the hottest in town, and this book is an excellent introduction to start engaging with it. But everyone should be keenly aware that mastering this technology comes with new responsibilities towards society. Let's progress the field with caution.

October 30, 2021   Max Welling

Preface

We live in a world where Artificial Intelligence (AI) has become a widely used term: there are movies about AI, journalists writing about AI, and CEOs talking about AI. Most importantly, there is AI in our daily lives, turning our phones, TVs, fridges, and vacuum cleaners into smartphones, smart TVs, smart fridges, and vacuum robots. We use AI, however, we still do not fully understand what "AI" is and how to formulate it, even though AI has been established as a separate field in the 1950s. Since then, many researchers pursue the holy grail of creating an artificial intelligence system that is capable of mimicking, understanding, and aiding humans through processing data and knowledge. In many cases, we have succeeded to outperform human beings on particular tasks in terms of speed and accuracy! Current AI methods do not necessarily imitate human processing (neither biologically nor cognitively) but rather are aimed at making a quick and accurate decision like navigating in cleaning a room or enhancing the quality of a displayed movie. In such tasks, *probability theory* is key since limited or poor quality of data or intrinsic behavior of a system forces us to quantify uncertainty. Moreover, *deep learning* has become a leading learning paradigm that allows learning hierarchical data representations. It draws its motivation from biological neural networks; however, the correspondence between deep learning and biological neurons is rather far-fetched. Nevertheless, deep learning has brought AI to the next level, achieving state-of-the-art performance in many decision-making tasks. The next step seems to be a combination of these two paradigms, probability theory and deep learning, to obtain powerful AI systems that are able to quantify their uncertainties about environments they operate in.

What Is This Book About Then? This book tackles the problem of formulating AI systems by combining probabilistic modeling and deep learning. Moreover, it goes beyond the typical predictive modeling and brings together supervised learning and unsupervised learning. The resulting paradigm, called *deep generative modeling*, utilizes the generative perspective on perceiving the surrounding world. It assumes that each phenomenon is driven by an underlying generative process that defines a joint distribution over random variables and their stochastic interactions, i.e., how events occur and in what order. The adjective "deep" comes from the fact that the distribution is parameterized using deep neural networks. There are two distinct traits of deep generative modeling. First, the application of deep neural networks allows rich and flexible parameterization of distributions. Second, the principled manner of modeling stochastic dependencies using probability theory ensures rigorous formulation and prevents potential flaws in reasoning. Moreover, probability theory provides a unified framework where the likelihood function plays a crucial role in quantifying uncertainty and defining objective functions.

Who Is This Book for Then? The book is designed to appeal to curious students, engineers, and researchers with a modest mathematical background in under-graduate calculus, linear algebra, probability theory, and the basics in machine learning, deep learning, and programming in Python and PyTorch (or other deep learning libraries). It should appeal to students and researchers from a variety of backgrounds, including computer science, engineering, data science, physics, and bioinformatics that wish to get familiar with deep generative modeling. In order to engage with a reader, the book introduces fundamental concepts with specific examples and code snippets. The full code accompanying the book is available online at:

https://github.com/jmtomczak/intro_dgm

The ultimate aim of the book is to outline the most important techniques in deep generative modeling and, eventually, enable readers to formulate new models and implement them.

The Structure of the Book The book consists of eight chapters that could be read separately and in (almost) any order. Chapter 1 introduces the topic and highlights important classes of deep generative models and general concepts. Chapters 2, 3 and 4 discuss modeling of *marginal* distributions while Chaps. 5, and 6 outline the material on modeling of *joint* distributions. Chapter 7 presents a class of latent variable models that are not learned through the likelihood-based objective. The last chapter, Chap. 8, indicates how deep generative modeling could be used in the fast-growing field of neural compression. All chapters are accompanied by code snippets to help understand how the presented methods could be implemented. The references are generally to indicate the original source of the presented material and provide further reading. Deep generating modeling is a broad field of study, and including all fantastic ideas is nearly impossible. Therefore, I would like to apologize for missing any paper. If anyone feels left out, it was not intentional from my side.

In the end, I would like to thank my wife, Ewelina, for her help and presence that gave me the strength to carry on with writing this book. I am also grateful to my parents for always supporting me, and my brother who spent a lot of time checking the first version of the book and the code.

Amsterdam, The Netherlands   Jakub M. Tomczak
November 1, 2021

Acknowledgments

This book, like many other books, would not have been possible without the contribution and help from many people. During my career, I was extremely privileged and lucky to work on deep generative modeling with an amazing set of people whom I would like to thank here (in alphabetical order): Tameem Adel, Rianne van den Berg, Taco Cohen, Tim Davidson, Nicola De Cao, Luka Falorsi, Eliseo Ferrante, Patrick Forré, Ioannis Gatopoulos, Efstratios Gavves, Adam Gonczarek, Amirhossein Habibian, Leonard Hasenclever, Emiel Hoogeboom, Maximilian Ilse, Thomas Kipf, Anna Kuzina, Christos Louizos, Yura Perugachi-Diaz, Ties van Rozendaal, Victor Satorras, Jerzy Świątek, Max Welling, Szymon Zaręba, and Maciej Zięba.

I would like to thank other colleagues with whom I worked on AI and had plenty of fascinating discussions (in alphabetical order): Davide Abati, Ilze Auzina, Babak Ehteshami Bejnordi, Erik Bekkers, Tijmen Blankevoort, Matteo De Carlo, Fuda van Diggelen, A.E. Eiben, Ali El Hassouni, Arkadiusz Gertych, Russ Greiner, Mark Hoogendoorn, Emile van Krieken, Gongjin Lan, Falko Lavitt, Romain Lepert, Jie Luo, ChangYong Oh, Siamak Ravanbakhsh, Diederik Roijers, David W. Romero, Annette ten Teije, Auke Wiggers, and Alessandro Zonta.

I am especially thankful to my brother, Kasper, who patiently read all sections, and ran and checked every single line of code in this book. You can't even imagine my gratitude for that!

I would like to thank my wife, Ewelina, for supporting me all the time and giving me the strength to finish this book. Without her help and understanding, it would be nearly impossible to accomplish this project. I would like to also express my gratitude to my parents, Elżbieta and Ryszard, for their support at different stages of my life because without them I would never be who I am now.

Contents

Chapter 1
Why Deep Generative Modeling?

1.1 AI Is Not Only About Decision Making

Before we start thinking about (deep) generative modeling, let us consider a simple example. Imagine we have trained a deep neural network that classifies images ($\mathbf{x} \in \mathbb{Z}^D$) of animals ($y \in \mathcal{Y}$, and $\mathcal{Y} = \{cat, dog, horse\}$). Further, let us assume that this neural network is trained really well so that it always classifies a proper class with a high probability $p(y|\mathbf{x})$. So far so good, right? The problem could occur though. As pointed out in [1], adding noise to images could result in completely false classification. An example of such a situation is presented in Fig. 1.1 where adding noise could shift predicted probabilities of labels; however, the image is barely changed (at least to us, human beings).

This example indicates that neural networks that are used to parameterize the conditional distribution $p(y|\mathbf{x})$ seem to lack semantic understanding of images. Further, we even hypothesize that learning *discriminative models* is not enough for proper decision making and creating AI. A machine learning system cannot rely on learning how to make a decision without *understanding* the reality and being able to express *uncertainty* about the surrounding world. How can we trust such a system if even a small amount of noise could change its internal beliefs and also shift its certainty from one decision to the other? How can we communicate with such a system if it is unable to properly express its opinion about whether its surrounding is new or not?

To motivate the importance of the concepts like *uncertainty* and *understanding* in decision making, let us consider a system that classifies objects, but this time into two classes: orange and blue. We assume we have some two-dimensional data (Fig. 1.2, left) and a new datapoint to be classified (a black cross in Fig. 1.2). We can make decisions using two approaches. First, a classifier could be formulated explicitly by modeling the conditional distribution $p(y|\mathbf{x})$ (Fig. 1.2, middle). Second, we can consider a joint distribution $p(\mathbf{x}, y)$ that could be further decomposed as $p(\mathbf{x}, y) = p(y|\mathbf{x}) \, p(\mathbf{x})$ (Fig. 1.2, right).

© The Author(s), under exclusive license to Springer Nature Switzerland AG 2022
J. M. Tomczak, *Deep Generative Modeling*,
https://doi.org/10.1007/978-3-030-93158-2_1

Fig. 1.1 An example of adding noise to an almost perfectly classified image that results in a shift of predicted label

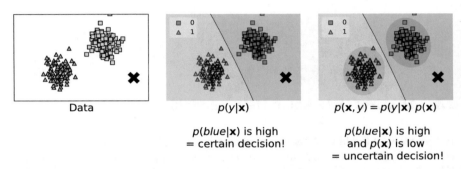

Fig. 1.2 And example of data (*left*) and two approaches to decision making: (*middle*) a discriminative approach and (*right*) a generative approach

After training a model using the discriminative approach, namely, the conditional distribution $p(y|\mathbf{x})$, we obtain a clear decision boundary. Then, we see that the black cross is farther away from the orange region; thus, the classifier assigns a higher probability to the blue label. As a result, the classifier is certain about the decision!

On the other hand, if we additionally fit a distribution $p(\mathbf{x})$, we observe that the black cross is not only farther away from the decision boundary, but also it is distant to the region where the blue datapoints lie. In other words, the black point is far away from the region of high probability mass. As a result, the (marginal) probability of the black cross $p(\mathbf{x} = black\ cross)$ is low, and the joint distribution $p(\mathbf{x} = black\ cross, y = blue)$ will be low as well and, thus, the decision is uncertain!

This simple example clearly indicates that if we want to build AI systems that make reliable decisions and can communicate with us, human beings, they must *understand* the environment first. For this purpose, they cannot simply learn how to make decisions, but they should be able to quantify their beliefs about their surrounding using the language of probability [2, 3]. In order to do that, we claim that estimating the distribution over objects, $p(\mathbf{x})$, is **crucial**.

From the *generative* perspective, knowing the distribution $p(\mathbf{x})$ is essential because:

- It could be used to assess whether a given object has been observed in the past or not.
- It could help to properly weight the decision.
- It could be used to assess uncertainty about the environment.

- It could be used to actively learn by interacting with the environment (e.g., by asking for labeling objects with low $p(\mathbf{x})$).
- And, eventually, it could be used to generate (synthesize) new objects.

Typically, in the literature of deep learning, *generative* models are treated as generators of new data. However, here we try to convey a new perspective where having $p(\mathbf{x})$ has much broader applicability, and this could be essential for building successful AI systems. Lastly, we would like to also make an obvious connection to *generative modeling* in machine learning, where formulating a proper *generative process* is crucial for understanding the phenomena of interest [3, 4]. However, in many cases, it is easier to focus on the other factorization, namely, $p(\mathbf{x}, y) = p(\mathbf{x}|y)\, p(y)$. We claim that considering $p(\mathbf{x}, y) = p(y|\mathbf{x})\, p(\mathbf{x})$ has clear advantages as mentioned before.

1.2 Where Can We Use (Deep) Generative Modeling?

With the development of neural networks and the increase in computational power, deep generative modeling has become one of the leading directions in AI. Its applications vary from typical modalities considered in machine learning, i.e., text analysis (e.g., [5]), image analysis (e.g., [6]), audio analysis (e.g., [7]), to problems in active learning (e.g., [8]), reinforcement learning (e.g., [9]), graph analysis (e.g., [10]), and medical imaging (e.g., [11]). In Fig. 1.3, we present graphically potential applications of deep generative modeling.

In some applications, it is indeed important to generate (synthesize) objects or modify features of objects to create new ones (e.g., an app turns a young person

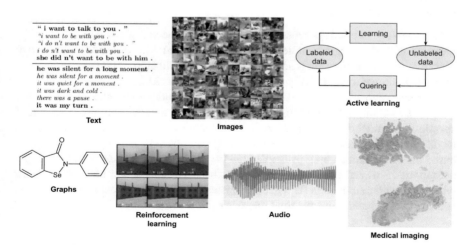

Fig. 1.3 Various potential applications of deep generative modeling

into an old one). However, in others like active learning it is important to ask for *uncertain* objects, i.e., objects with low $p(\mathbf{x})$) that should be labeled by an oracle. In reinforcement learning, on the other hand, generating the next most likely situation (states) is crucial for taking actions by an agent. For medical applications, explaining a decision, e.g., in terms of the probability of the label **and** the object, is definitely more informative to a human doctor than simply assisting with a diagnosis label. If an AI system would be able to indicate how certain it is and also quantify whether the object is suspicious (i.e., low $p(\mathbf{x})$) or not, then it might be used as an independent specialist that outlines its own opinion.

These examples clearly indicate that many fields, if not all, could highly benefit from (deep) generative modeling. Obviously, there are many mechanisms that AI systems should be equipped with. However, we claim that the generative modeling capability is definitely one of the most important ones, as outlined in the above-mentioned cases.

1.3   How to Formulate (Deep) Generative Modeling?

At this point, after highlighting the importance and wide applicability of (deep) generative modeling, we should ask ourselves how to formulate (deep) generative models. In other words, how to express $p(\mathbf{x})$ that we mentioned already multiple times.

We can divide (deep) generative modeling into four main groups (see Fig. 1.4):

- Autoregressive generative models (ARM)
- Flow-based models
- Latent variable models
- Energy-based models

We use *deep* in brackets because most of what we have discussed so far could be modeled without using neural networks. However, neural networks are flexible and powerful and, therefore, they are widely used to parameterize generative models. From now on, we focus entirely on deep generative models.

As a side note, please treat this taxonomy as a guideline that helps us to navigate through this book, not something written in stone. Personally, I am not a big fan of spending too much time on categorizing and labeling science because it very often results in antagonizing and gatekeeping. Anyway, there is also a group of models based on the score matching principle [12–14] that do not necessarily fit our simple taxonomy. However, as pointed out in [14], these models share a lot of similarities with latent variable models (if we treat consecutive steps of a stochastic process as latent variables) and, thus, we treat them as such.

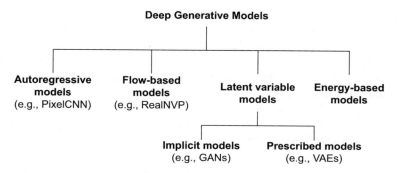

Fig. 1.4 A taxonomy of deep generative models

1.3.1 Autoregressive Models

The first group of deep generative models utilizes the idea of **autoregressive modeling** (ARM). In other words, the distribution over **x** is represented in an autoregressive manner:

$$p(\mathbf{x}) = p(x_0) \prod_{i=1}^{D} p(x_i|\mathbf{x}_{<i}), \tag{1.1}$$

where $\mathbf{x}_{<i}$ denotes all **x**'s up to the index i.

Modeling all conditional distributions $p(x_i|\mathbf{x}_{<i})$ would be computationally inefficient. However, we can take advantage of *causal convolutions* as presented in [7] for audio and in [15, 16] for images. We will discuss ARMs more in depth in Chap. 2.

1.3.2 Flow-Based Models

The change of variables formula provides a principled manner of expressing a density of a random variable by transforming it with an invertible transformation f [17]:

$$p(\mathbf{x}) = p\big(\mathbf{z} = f(\mathbf{x})\big)|\mathbf{J}_{f(\mathbf{x})}|, \tag{1.2}$$

where $\mathbf{J}_{f(\mathbf{x})}$ denotes the Jacobian matrix.

We can parameterize f using deep neural networks; however, it cannot be any arbitrary neural networks, because we must be able to calculate the Jacobian matrix. First ideas of using the change of variable formulate focused on linear, volume-preserving transformations that yields $|\mathbf{J}_{f(\mathbf{x})}| = 1$ [18, 19]. Further attempts utilized theorems on matrix determinants that resulted in specific non-linear transformations, namely, planar flows [20] and Sylvester flows [21, 22]. A different approach focuses on formulating invertible transformations for which the Jacobian-determinant could be calculated easily like for coupling layers in RealNVP [23]. Recently, arbitrary neural networks are constrained in such a way they are invertible and the Jacobian-determinant is approximated [24–26].

In the case of the discrete distributions (e.g., integers), for the probability mass functions, there is no change of volume and, therefore, the change of variables formula takes the following form:

$$p(\mathbf{x}) = p(\mathbf{z} = f(\mathbf{x})). \tag{1.3}$$

Integer discrete flows propose to use affine coupling layers with rounding operators to ensure the integer-valued output [27]. A generalization of the affine coupling layer was further investigated in [28].

All generative models that take advantage of the change of variables formula are referred to as **flow-based models** or *flows* for short. We will discuss flows in Chap. 3.

1.3.3   Latent Variable Models

The idea behind **latent variable models** is to assume a lower-dimensional latent space and the following generative process:

$$\mathbf{z} \sim p(\mathbf{z})$$

$$\mathbf{x} \sim p(\mathbf{x}|\mathbf{z}).$$

In other words, the latent variables correspond to hidden factors in data, and the conditional distribution $p(\mathbf{x}|\mathbf{z})$ could be treated as a *generator*.

The most widely known latent variable model is the **probabilistic Principal Component Analysis** (pPCA) [29] where $p(\mathbf{z})$ and $p(\mathbf{x}|\mathbf{z})$ are Gaussian distributions, and the dependency between $\mathbf{z}$ and $\mathbf{x}$ is linear.

A non-linear extension of the pPCA with arbitrary distributions is the **Variational Auto-Encoder** (VAE) framework [30, 31]. To make the inference tractable, variational inference is utilized to approximate the posterior $p(\mathbf{z}|\mathbf{x})$, and neural networks are used to parameterize the distributions. Since the publication of the seminal papers by Kingma and Welling [30], Rezende et al. [31], there were multiple extensions of this framework, including working on more powerful variational posteriors [19, 21, 22, 32], priors [33, 34], and decoders [35]. Interesting directions include considering different topologies of the latent space, e.g., the hyperspherical latent space [36]. In VAEs and the pPCA all distributions must be defined upfront and, therefore, they are called *prescribed models*. We will pay special attention to this group of deep generative models in Chap. 4.

So far, ARMs, flows, the pPCA, and VAEs are probabilistic models with the objective function being the *log-likelihood function* that is closely related to using the Kullback–Leibler divergence between the data distribution and the model distribution. A different approach utilizes an *adversarial loss* in which a discriminator $D(\cdot)$ determines a difference between real data and synthetic data provided by a generator in the implicit form, namely, $p(\mathbf{x}|\mathbf{z}) = \delta(\mathbf{x} - G(\mathbf{z}))$, where $\delta(\cdot)$ is the Dirac delta. This group of models is called *implicit models*, and Generative Adversarial Networks (GANs) [6] become one of the first successful deep generative models for synthesizing realistic-looking objects (e.g., images). See Chap. 7 for more details.

1.3.4 Energy-Based Models

Physics provide an interesting perspective on defining a group of generative models through defining an *energy function*, $E(\mathbf{x})$, and, eventually, the Boltzmann distribution:

$$p(\mathbf{x}) = \frac{\exp\{-E(\mathbf{x})\}}{Z}, \tag{1.4}$$

where $Z = \sum_{\mathbf{x}} \exp\{-E(\mathbf{x})\}$ is the partition function.

In other words, the distribution is defined by the exponentiated energy function that is further normalized to obtain values between 0 and 1 (i.e., probabilities). There is much more into that if we think about physics, but we do not require delving into that. I refer to [37] as a great starting point for that.

Models defined by an energy function are referred to as *energy-based models* (EBMs) [38]. The main idea behind EBMs is to formulate the energy function and calculate (or rather approximate) the partition function. The largest group of EBMs consists of *Boltzmann Machines* that entangle $\mathbf{x}$'s through a bilinear form, i.e., $E(\mathbf{x}) = \mathbf{x}^\top \mathbf{W} \mathbf{x}$ [39, 40]. Introducing latent variables and taking $E(\mathbf{x}, \mathbf{z}) = \mathbf{x}^\top \mathbf{W} \mathbf{z}$ results in *Restricted Boltzmann Machines* [41]. The idea of Boltzmann machines could be further extended to the joint distribution over $\mathbf{x}$ and y as it is done, e.g., in classification Restricted Boltzmann Machines [42]. Recently, it has been shown that an arbitrary neural network could be used to define the joint distribution [43]. We will discuss how this could be accomplished in Chap. 6.

1.3.5   Overview

In Table 1.1, we compared all four groups of models (with a distinction between implicit latent variable models and prescribed latent variable models) using arbitrary criteria like:

- Whether training is typically stable
- Whether it is possible to calculate the likelihood function
- Whether one can use a model for lossy or lossless compression
- Whether a model could be used for representation learning

All likelihood-based models (i.e., ARMs, flows, EBMs, and prescribed models like VAEs) can be trained in a stable manner, while implicit models like GANs suffer from instabilities. In the case of the non-linear prescribed models like VAEs, we must remember that the likelihood function cannot be exactly calculated, and only a lower-bound could be provided. Similarly, EBMs require calculating the partition function that is analytically intractable problem. As a result, we can get the unnormalized probability or an approximation at best. ARMs constitute one of the best likelihood-based models; however, their sampling process is extremely slow due to the autoregressive manner of generating new content. EBMs require running a Monte Carlo method to receive a sample. Since we operate on high-dimensional objects, this is a great obstacle for using EBMs widely in practice. All other approaches are relatively fast. In the case of compression, VAEs are models that allow us to use a bottleneck (the latent space). On the other hand, ARMs and flows could be used for lossless compression since they are density estimators and provide the exact likelihood value. Implicit models cannot be directly used for compression; however, recent works use GANs to improve image compression [44]. Flows, prescribed models, and EBMs (if use latents) could be used for representation learning, namely, learning a set of random variables that summarize data in some way and/or disentangle factors in data. The question about what is a good representation is a different story and we refer a curious reader to literature, e.g., [45].

Table 1.1 A comparison of deep generative models

Generative models	Training	Likelihood	Sampling	Compression	Representation
Autoregressive models	Stable	Exact	Slow	Lossless	No
Flow-based models	Stable	Exact	Fast/slow	Lossless	Yes
Implicit models	Unstable	No	Fast	No	No
Prescribed models	Stable	Approximate	Fast	Lossy	Yes
Energy-based models	Stable	Unnormalized	Slow	Rather not	Yes

1.4 Purpose and Content of This Book

This book is intended as an introduction to the field of deep generative modeling. Its goal is to convince you, dear reader, to the philosophy of generative modeling and show you its beauty! Deep generative modeling is an interesting hybrid that combines probability theory, statistics, probabilistic machine learning, and deep learning in a single framework. However, to be able to follow the ideas presented in this book it is advised to possess knowledge in algebra and calculus, probability theory and statistics, the basics of machine learning and deep learning, and programming with Python. Knowing PyTorch[1] is highly recommended since all code snippets are written in PyTorch. However, knowing other deep learning frameworks like Keras, Tensorflow, or JAX should be sufficient to understand the code.

In this book, we will not review machine learning concepts or building blocks in deep learning unless it is essential to comprehend a given topic. Instead, we will delve into models and training algorithms of deep generative models. We will either discuss the marginal models, such as autoregressive models (Chap. 2), flow-based models (Chap. 3): RealNVP, Integer Discrete Flows, and residual and DenseNet flows, latent variable models (Chap. 4): Variational Auto-Encoder and its components, hierarchical VAEs, and Diffusion-based deep generative models, or frameworks for modeling the joint distribution like hybrid modeling (Chap. 5) and energy-based models (Chap. 6). Eventually, we will present how deep generative modeling could be useful for data compression within the neural compression framework (Chap. 8). In general, the book is organized in such a way that each chapter could be followed independently from the others and in an order that suits a reader best.

So who is the target audience of this book? Well, hopefully everybody who is interested in AI, but there are two groups who could definitely benefit from the presented content. The first target audience is university students who want to go beyond standard courses on machine learning and deep learning. The second group is research engineers who want to broaden their knowledge on AI or prefer to make the next step in their careers and learn about the next generation of AI systems. Either way, the book is intended for curious minds who want to understand AI and learn not only about theory but also how to implement the discussed material. For this purpose, each topic is associated with general discussion and introduction that is further followed by formal formulations and a piece of code (in PyTorch). The intention of this book is to truly understand deep generative modeling that, in the humble opinion of the author of this book, is only possible if one can not only derive a model but also implement it. Therefore, this book is accompanied by the following code repository:

https://github.com/jmtomczak/intro_dgm

---

[1] https://pytorch.org/.

References

1. Christian Szegedy, Wojciech Zaremba, Ilya Sutskever, Joan Bruna, Dumitru Erhan, Ian Goodfellow, and Rob Fergus. Intriguing properties of neural networks. In *2nd International Conference on Learning Representations, ICLR 2014*, 2014.
2. Christopher M Bishop. Model-based machine learning. *Philosophical Transactions of the Royal Society A: Mathematical, Physical and Engineering Sciences*, 371(1984):20120222, 2013.
3. Zoubin Ghahramani. Probabilistic machine learning and artificial intelligence. *Nature*, 521(7553):452–459, 2015.
4. Julia A Lasserre, Christopher M Bishop, and Thomas P Minka. Principled hybrids of generative and discriminative models. In *2006 IEEE Computer Society Conference on Computer Vision and Pattern Recognition (CVPR'06)*, volume 1, pages 87–94. IEEE, 2006.
5. Samuel Bowman, Luke Vilnis, Oriol Vinyals, Andrew Dai, Rafal Jozefowicz, and Samy Bengio. Generating sentences from a continuous space. In *Proceedings of The 20th SIGNLL Conference on Computational Natural Language Learning*, pages 10–21, 2016.
6. Ian J Goodfellow, Jean Pouget-Abadie, Mehdi Mirza, Bing Xu, David Warde-Farley, Sherjil Ozair, Aaron Courville, and Yoshua Bengio. Generative adversarial networks. *arXiv preprint arXiv:1406.2661*, 2014.
7. Aaron van den Oord, Sander Dieleman, Heiga Zen, Karen Simonyan, Oriol Vinyals, Alex Graves, Nal Kalchbrenner, Andrew Senior, and Koray Kavukcuoglu. WaveNet: A generative model for raw audio. *arXiv preprint arXiv:1609.03499*, 2016.
8. Samarth Sinha, Sayna Ebrahimi, and Trevor Darrell. Variational adversarial active learning. In *Proceedings of the IEEE/CVF International Conference on Computer Vision*, pages 5972–5981, 2019.
9. David Ha and Jürgen Schmidhuber. World models. *arXiv preprint arXiv:1803.10122*, 2018.
10. GraphVAE: Towards generation of small graphs using variational autoencoders, author=Simonovsky, Martin and Komodakis, Nikos, booktitle=International Conference on Artificial Neural Networks, pages=412–422, year=2018, organization=Springer.
11. Maximilian Ilse, Jakub M Tomczak, Christos Louizos, and Max Welling. DIVA: Domain invariant variational autoencoders. In *Medical Imaging with Deep Learning*, pages 322–348. PMLR, 2020.
12. Aapo Hyvärinen and Peter Dayan. Estimation of non-normalized statistical models by score matching. *Journal of Machine Learning Research*, 6(4), 2005.
13. Yang Song and Stefano Ermon. Generative modeling by estimating gradients of the data distribution. *arXiv preprint arXiv:1907.05600*, 2019.
14. Yang Song, Jascha Sohl-Dickstein, Diederik P Kingma, Abhishek Kumar, Stefano Ermon, and Ben Poole. Score-based generative modeling through stochastic differential equations. In *International Conference on Learning Representations*, 2020.
15. Aaron Van Oord, Nal Kalchbrenner, and Koray Kavukcuoglu. Pixel recurrent neural networks. In *International Conference on Machine Learning*, pages 1747–1756. PMLR, 2016.
16. Aäron van den Oord, Nal Kalchbrenner, Oriol Vinyals, Lasse Espeholt, Alex Graves, and Koray Kavukcuoglu. Conditional image generation with PixelCNN decoders. In *Proceedings of the 30th International Conference on Neural Information Processing Systems*, pages 4797–4805, 2016.
17. Oren Rippel and Ryan Prescott Adams. High-dimensional probability estimation with deep density models. *arXiv preprint arXiv:1302.5125*, 2013.
18. Laurent Dinh, David Krueger, and Yoshua Bengio. NICE: Non-linear independent components estimation. *arXiv preprint arXiv:1410.8516*, 2014.
19. Jakub M Tomczak and Max Welling. Improving variational auto-encoders using householder flow. *arXiv preprint arXiv:1611.09630*, 2016.
20. Danilo Rezende and Shakir Mohamed. Variational inference with normalizing flows. In *International Conference on Machine Learning*, pages 1530–1538. PMLR, 2015.

21. Rianne Van Den Berg, Leonard Hasenclever, Jakub M Tomczak, and Max Welling. Sylvester normalizing flows for variational inference. In *34th Conference on Uncertainty in Artificial Intelligence 2018, UAI 2018*, pages 393–402. Association For Uncertainty in Artificial Intelligence (AUAI), 2018.
22. Emiel Hoogeboom, Victor Garcia Satorras, Jakub M Tomczak, and Max Welling. The convolution exponential and generalized Sylvester flows. *arXiv preprint arXiv:2006.01910*, 2020.
23. Laurent Dinh, Jascha Sohl-Dickstein, and Samy Bengio. Density estimation using Real NVP. *arXiv preprint arXiv:1605.08803*, 2016.
24. Jens Behrmann, Will Grathwohl, Ricky TQ Chen, David Duvenaud, and Jörn-Henrik Jacobsen. Invertible residual networks. In *International Conference on Machine Learning*, pages 573–582. PMLR, 2019.
25. Ricky TQ Chen, Jens Behrmann, David Duvenaud, and Jörn-Henrik Jacobsen. Residual flows for invertible generative modeling. *arXiv preprint arXiv:1906.02735*, 2019.
26. Yura Perugachi-Diaz, Jakub M Tomczak, and Sandjai Bhulai. Invertible DenseNets with Concatenated LipSwish. *Advances in Neural Information Processing Systems*, 2021.
27. Emiel Hoogeboom, Jorn WT Peters, Rianne van den Berg, and Max Welling. Integer discrete flows and lossless compression. *arXiv preprint arXiv:1905.07376*, 2019.
28. Jakub M Tomczak. General invertible transformations for flow-based generative modeling. *INNF+*, 2021.
29. Michael E Tipping and Christopher M Bishop. Probabilistic principal component analysis. *Journal of the Royal Statistical Society: Series B (Statistical Methodology)*, 61(3):611–622, 1999.
30. Diederik P Kingma and Max Welling. Auto-encoding variational Bayes. *arXiv preprint arXiv:1312.6114*, 2013.
31. Danilo Jimenez Rezende, Shakir Mohamed, and Daan Wierstra. Stochastic backpropagation and approximate inference in deep generative models. In *International conference on machine learning*, pages 1278–1286. PMLR, 2014.
32. Durk P Kingma, Tim Salimans, Rafal Jozefowicz, Xi Chen, Ilya Sutskever, and Max Welling. Improved variational inference with inverse autoregressive flow. *Advances in Neural Information Processing Systems*, 29:4743–4751, 2016.
33. Xi Chen, Diederik P Kingma, Tim Salimans, Yan Duan, Prafulla Dhariwal, John Schulman, Ilya Sutskever, and Pieter Abbeel. Variational lossy autoencoder. *arXiv preprint arXiv:1611.02731*, 2016.
34. Jakub Tomczak and Max Welling. VAE with a VampPrior. In *International Conference on Artificial Intelligence and Statistics*, pages 1214–1223. PMLR, 2018.
35. Ishaan Gulrajani, Kundan Kumar, Faruk Ahmed, Adrien Ali Taiga, Francesco Visin, David Vazquez, and Aaron Courville. PixelVAE: A latent variable model for natural images. *arXiv preprint arXiv:1611.05013*, 2016.
36. Tim R Davidson, Luca Falorsi, Nicola De Cao, Thomas Kipf, and Jakub M Tomczak. Hyperspherical variational auto-encoders. In *34th Conference on Uncertainty in Artificial Intelligence 2018, UAI 2018*, pages 856–865. Association For Uncertainty in Artificial Intelligence (AUAI), 2018.
37. Edwin T Jaynes. *Probability theory: The logic of science*. Cambridge university press, 2003.
38. Yann LeCun, Sumit Chopra, Raia Hadsell, M Ranzato, and F Huang. A tutorial on energy-based learning. *Predicting structured data*, 1(0), 2006.
39. David H Ackley, Geoffrey E Hinton, and Terrence J Sejnowski. A learning algorithm for Boltzmann machines. *Cognitive science*, 9(1):147–169, 1985.
40. Geoffrey E Hinton, Terrence J Sejnowski, et al. Learning and relearning in Boltzmann machines. *Parallel distributed processing: Explorations in the microstructure of cognition*, 1(282-317):2, 1986.
41. Geoffrey E Hinton. A practical guide to training restricted Boltzmann machines. In *Neural networks: Tricks of the trade*, pages 599–619. Springer, 2012.

42. Hugo Larochelle and Yoshua Bengio. Classification using discriminative restricted Boltzmann machines. In *Proceedings of the 25th international conference on Machine learning*, pages 536–543, 2008.
43. Will Grathwohl, Kuan-Chieh Wang, Joern-Henrik Jacobsen, David Duvenaud, Mohammad Norouzi, and Kevin Swersky. Your classifier is secretly an energy based model and you should treat it like one. In *International Conference on Learning Representations*, 2019.
44. Fabian Mentzer, George D Toderici, Michael Tschannen, and Eirikur Agustsson. High-fidelity generative image compression. *Advances in Neural Information Processing Systems*, 33, 2020.
45. Yoshua Bengio, Aaron Courville, and Pascal Vincent. Representation learning: A review and new perspectives. *IEEE transactions on pattern analysis and machine intelligence*, 35(8):1798–1828, 2013.

Chapter 2
Autoregressive Models

2.1 Introduction

Before we start discussing how we can model the distribution $p(\mathbf{x})$, we refresh our memory about the core rules of probability theory, namely, the **sum rule** and the **product rule**. Let us introduce two random variables $\mathbf{x}$ and $\mathbf{y}$. Their joint distribution is $p(\mathbf{x}, \mathbf{y})$. The **product rule** allows us to *factorize* the joint distribution in two manners, namely:

$$p(\mathbf{x}, \mathbf{y}) = p(\mathbf{x}|\mathbf{y})p(\mathbf{y}) \tag{2.1}$$

$$= p(\mathbf{y}|\mathbf{x})p(\mathbf{x}). \tag{2.2}$$

In other words, the joint distribution could be represented as a product of a marginal distribution and a conditional distribution. The **sum rule** tells us that if we want to calculate the marginal distribution over one of the variables, we must integrate out (or sum out) the other variable, that is:

$$p(\mathbf{x}) = \sum_{\mathbf{y}} p(\mathbf{x}, \mathbf{y}). \tag{2.3}$$

These two rules will play a crucial role in probability theory and statistics and, in particular, in formulating deep generative models.

Now, let us consider a high-dimensional random variable $\mathbf{x} \in \mathcal{X}^D$ where $\mathcal{X} = \{0, 1, \ldots, 255\}$ (e.g., pixel values) or $\mathcal{X} = \mathbb{R}$. Our goal is to model $p(\mathbf{x})$. Before we jump into thinking of specific parameterization, let us first apply the product rule to express the joint distribution in a different manner:

$$p(\mathbf{x}) = p(x_1) \prod_{d=2}^{D} p(x_d|\mathbf{x}_{<d}), \tag{2.4}$$

© The Author(s), under exclusive license to Springer Nature Switzerland AG 2022
J. M. Tomczak, *Deep Generative Modeling*,
https://doi.org/10.1007/978-3-030-93158-2_2

where $\mathbf{x}_{<d} = [x_1, x_2, \ldots, x_{d-1}]^\top$. For instance, for $\mathbf{x} = [x_1, x_2, x_3]^\top$, we have $p(\mathbf{x}) = p(x_1)p(x_2|x_1)p(x_3|x_1, x_2)$.

As we can see, the product rule applied multiple times to the joint distribution provides a principled manner of factorizing the joint distribution into many conditional distributions. That's great news! However, modeling all conditional distributions $p(x_d|\mathbf{x}_{<d})$ separately is simply infeasible! If we did that, we would obtain D separate models, and the complexity of each model would grow due to varying conditioning. A natural question is whether we can do better, and the answer is yes.

2.2   Autoregressive Models Parameterized by Neural Networks

As mentioned earlier, we aim for modeling the joint distribution $p(\mathbf{x})$ using conditional distributions. A potential solution to the issue of using D separate model is utilizing a single, shared model for the conditional distribution. However, we need to make some assumptions to use such a shared model. In other words, we look for an **autoregressive model** (ARM). In the next subsection, we outline ARMs parameterized with various *neural networks*. After all, we are talking about deep generative models so using a neural network is not surprising, isn't it?

2.2.1   Finite Memory

The first attempt to limiting the complexity of a conditional model is to assume a *finite memory*. For instance, we can assume that each variable is dependent on no more than two other variables, namely:

$$p(\mathbf{x}) = p(x_1)p(x_2|x_1) \prod_{d=3}^{D} p(x_d|x_{d-1}, x_{d-2}). \tag{2.5}$$

Then, we can use a small neural network, e.g., multi-layered perceptron (MLP), to predict the distribution of x_d. If $\mathcal{X} = \{0, 1, \ldots, 255\}$, the MLP takes x_{d-1}, x_{d-2} and outputs probabilities for the categorical distribution of x_d, θ_d. The MLP could be of the following form:

$$[x_{d-1}, x_{d-2}] \rightarrow \text{Linear}(2, M) \rightarrow \text{ReLU} \rightarrow \text{Linear}(M, 256) \rightarrow \text{softmax} \rightarrow \theta_d, \tag{2.6}$$

where M denotes the number of hidden units, e.g., $M = 300$. An example of this approach is depicted in Fig. 2.1.

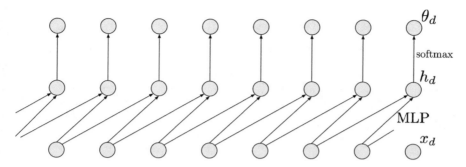

Fig. 2.1 An example of applying a shared MLP depending on two last inputs. Inputs are denoted by blue nodes (bottom), intermediate representations are denoted by orange nodes (middle), and output probabilities are denoted by green nodes (top). Notice that a probability θ_d is **not** dependent on x_d

It is important to notice that now we use a single, shared MLP to predict probabilities for x_d. Such a model is not only non-linear but also its parameterization is convenient due to a relatively small number of weights to be trained. However, the obvious drawback of this approach is a **limited memory** (i.e., only two last variables in our example). Moreover, it is unclear a priori how many variables we should use in conditioning. In many problems, e.g., image processing, learning *long-range statistics* is crucial to understand complex patterns in data; therefore, having long-range memory is essential.

2.2.2 Long-Range Memory Through RNNs

A possible solution to the problem of a short-range memory modeled by an MLP relies on applying a recurrent neural network (RNN) [1, 2]. In other words, we can model the conditional distributions as follows [3]:

$$p(x_d|\mathbf{x}_{<d}) = p\left(x_d|\text{RNN}(x_{d-1}, h_{d-1})\right), \tag{2.7}$$

where $h_d = \text{RNN}(x_{d-1}, h_{d-1})$, and h_d is a hidden context, which acts as a *memory* that allows learning long-range dependencies. An example of using an RNN is presented in Fig. 2.2.

This approach gives a single parameterization, thus, it is efficient and also solves the problem of a finite memory. So far so good! Unfortunately, RNNs suffer from other issues, namely:

- They are sequential, hence, slow.

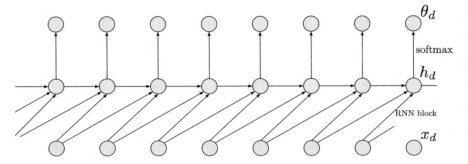

Fig. 2.2 An example of applying an RNN depending on two last inputs. Inputs are denoted by blue nodes (bottom), intermediate representations are denoted by orange nodes (middle), and output probabilities are denoted by green nodes (top). Notice that compared to the approach with a shared MLP, there is an additional dependency between intermediate nodes h_d

- If they are badly conditioned (i.e., the eigenvalues of a weight matrix are larger or smaller than 1, then they suffer from exploding or vanishing gradients, respectively, that hinders learning long-range dependencies.

There exist methods to help training RNNs like gradient clipping or, more generally, gradient regularization [4] or orthogonal weights [5]. However, here we are not interested in looking into rather specific solutions to new problems. We seek for a different parameterization that could solve our original problem, namely, modeling long-range dependencies in an ARM.

2.2.3   Long-Range Memory Through Convolutional Nets

In [6, 7] it was noticed that convolutional neural networks (CNNs) could be used instead of RNNs to model long-range dependencies. To be more precise, one-dimensional convolutional layers (Conv1D) could be stacked together to process sequential data. The advantages of such an approach are the following:

- Kernels are shared (i.e., an efficient parameterization).
- The processing is done in parallel that greatly speeds up computations.
- By stacking more layers, the effective kernel size grows with the network depth.

These three traits seem to place Conv1D-based neural networks as a perfect solution to our problem. However, can we indeed use them straight away?

A Conv1D can be applied to calculate embeddings like in [7], but it cannot be used for autoregressive models. Why? Because we need convolutions to be **causal** [8]. *Causal* in this context means that a Conv1D layer is dependent on the last k inputs but the current one (*option A*) or with the current one (*option B*). In other words, we must "cut" the kernel in half and forbid it to look into the next variables

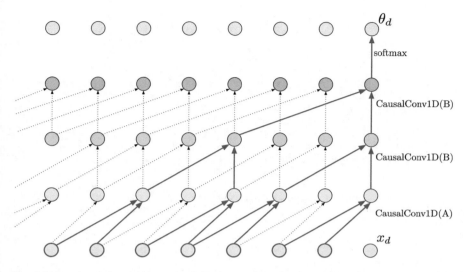

Fig. 2.3 An example of applying causal convolutions. The kernel size is 2, but by applying dilation in higher layers, a much larger input could be processed (red edges), thus, a larger memory is utilized. Notice that the first layers must be option A to ensure proper processing

(look into the future). Importantly, the option A is required in the first layer because the final output (i.e., the probabilities θ_d) cannot be dependent on x_d. Additionally, if we are concerned about the effective kernel size, we can use *dilation* larger than 1.

In Fig. 2.3 we present an example of a neural network consisting of 3 *causal Conv1D* layers. The first CausalConv1D is of type A, i.e., it does not take into account only the last k inputs without the current one. Then, in the next two layers, we use CausalConv1D (option B) with dilations 2 and 3. Typically, the dilation values are 1, 2, 4, and 8 (v.d. Oord et al., 2016a); however, taking 2 and 4 would not nicely fit in a figure. We highlight in red all connections that go from the output layer to the input layer. As we can notice, stacking CausalConv1D layers with the dilation larger than 1 allows us to learn long-range dependencies (in this example, by looking at 7 last inputs).

An example of an implementation of CausalConv1D layer is presented below. If you are still confused about the option A and the option B, please analyze the code snippet step-by-step.

```
class CausalConv1d(nn.Module):
    def __init__(self, in_channels, out_channels, kernel_size,
    dilation, A=False, **kwargs):
        super(CausalConv1d, self).__init__()

        # The general idea is the following: We take the built-in
    PyTorch Conv1D. Then, we must pick a proper padding, because
    we must ensure the convolutional is causal. Eventually, we
```

must remove some final elements of the output, because we simply don't need them! Since CausalConv1D is still a convolution, we must define the kernel size, dilation, and whether it is option A (A=True) or option B (A=False). Remember that by playing with dilation we can enlarge the size of the memory.

```
 6
 7        # attributes:
 8        self.kernel_size = kernel_size
 9        self.dilation = dilation
10        self.A = A # whether option A (A=True) or B (A=False)
11        self.padding = (kernel_size - 1) * dilation + A * 1
12
13        # we will do padding by ourselves in the forward pass!
14        self.conv1d = torch.nn.Conv1d(in_channels, out_channels,
15                                      kernel_size, stride=1,
16                                      padding=0,
17                                      dilation=dilation,**kwargs)
18
19    def forward(self, x):
20        # We do padding only from the left! This is more
    efficient implementation.
21        x = torch.nn.functional.pad(x, (self.padding, 0))
22        conv1d_out = self.conv1d(x)
23        if self.A:
24            # Remember, we cannot be dependent on the current
    component; therefore, the last element is removed.
25            return conv1d_out[:, :, : -1]
26        else:
27            return conv1d_out
```

Listing 2.1 Causal convolution 1D

The CausalConv1D layers are better-suited to modeling sequential data than RNNs. They obtain not only better results (e.g., classification accuracy) but also allow learning long-range dependencies more efficiently than RNNs [8]. Moreover, they do not suffer from exploding/vanishing gradient issues. As a result, they seem to be a perfect parameterization for autoregressive models! Their supremacy has been proven in many cases, including audio processing by WaveNet, a neural network consisting of CausalConv1D layers [9], or image processing by PixelCNN, a model with CausalConv2D components [10].

Then, is there any drawback of applying autoregressive models parameterized by causal convolutions? Unfortunately, yes, there is and it is connected with sampling. If we want to evaluate probabilities for given inputs, we need to calculate the forward pass where all calculations are done in parallel. However, if we want to sample new objects, we must iterate through all positions (think of a big for-loop, from the first variable to the last one) and iteratively predict probabilities and sample new values. Since we use convolutions to parameterize the model, we must do D full forward passes to get the final sample. That is a big waste, but, unfortunately, that is the price we must pay for all "goodies" following from the convolutional-based parameterization of the ARM. Fortunately, there is on-going research on speeding up computations, e.g., see [11].

2.3 Deep Generative Autoregressive Model in Action!

Alright, let us talk more about details and how to implement an ARM. Here, and in the whole book, we focus on images, e.g., $\mathbf{x} \in \{0, 1, \ldots, 15\}^{64}$. Since images are represented by integers, we will use the categorical distribution to represent them (in next chapters, we will comment on the choice of distribution for images and present some alternatives). We model $p(\mathbf{x})$ using an ARM parameterized by CausalConv1D layers. As a result, each conditional is the following:

$$p(x_d | \mathbf{x}_{<d}) = \text{Categorical} \left(x_d | \theta_d \left(\mathbf{x}_{<d} \right) \right) \tag{2.8}$$

$$= \prod_{l=1}^{L} \left(\theta_{d,l} \right)^{[x_d = l]}, \tag{2.9}$$

where $[a = b]$ is the Iverson bracket (i.e., $[a = b] = 1$ if $a = b$, and $[a = b] = 0$ if $a \neq b$), and $\theta_d \left(\mathbf{x}_{<d} \right) \in [0, 1]^{16}$ is the output of the CausalConv1D-based neural network with the softmax in the last layer, so $\sum_{l=1}^{L} \theta_{d,l} = 1$. To be very clear, the last layer must have 16 output channels (because there are 16 possible values per pixel), and the softmax is taken over these 16 values. We stack CausalConv1D layers with non-linear activation functions in between (e.g., LeakyRELU). Of course, we must remember about taking the option A CausalConv1D as the first layer! Otherwise we break the assumption about taking into account x_d in predicting θ_d.

What about the objective function? ARMs are the likelihood-based models, so for given N i.i.d. datapoints $\mathcal{D} = \{\mathbf{x}_1, \ldots, \mathbf{x}_N\}$, we aim at maximizing the logarithm of the likelihood function, that is (we will use the product and sum rules again):

$$\ln p(\mathcal{D}) = \ln \prod_n p(\mathbf{x}_n) \tag{2.10}$$

$$= \sum_n \ln p(\mathbf{x}_n) \tag{2.11}$$

$$= \sum_n \ln \prod_d p(x_{n,d} | \mathbf{x}_{n,<d}) \tag{2.12}$$

$$= \sum_n \left(\sum_d \ln p(x_{n,d} | \mathbf{x}_{n,<d}) \right) \tag{2.13}$$

$$= \sum_n \left(\sum_d \ln \text{Categorical} \left(x_d | \theta_d \left(\mathbf{x}_{<d} \right) \right) \right) \tag{2.14}$$

$$= \sum_n \left(\sum_d \left(\sum_{l=1}^{L} [x_d = l] \ln \theta_d \left(\mathbf{x}_{<d} \right) \right) \right). \tag{2.15}$$

For simplicity, we assumed that $\mathbf{x}_{<1} = \emptyset$, i.e., no conditioning. As we can notice, the objective function takes a very nice form! First, the logarithm over the i.i.d. data $\mathcal{D}$ results in a sum over datapoints of the logarithm of individual distributions $p(\mathbf{x}_n)$. Second, applying the product rule, together with the logarithm, results in another sum, this time over dimensions. Eventually, by parameterizing the conditionals by CausalConv1D, we can calculate all θ_d in one forward pass and then check the pixel value (see the last line of $\ln p(\mathcal{D})$). Ideally, we want $\theta_{d,l}$ to be as close to 1 as possible if $x_d = l$.

2.3.1   Code

Uff... Alright, let's take a look at some code. The full code is available under the following: https://github.com/jmtomczak/intro_dgm. Here, we focus only on the code for the model. We provide details in the comments.

```
 1  class ARM(nn.Module):
 2      def __init__(self, net, D=2, num_vals=256):
 3          super(ARM, self).__init__()
 4
 5          # Remember, always credit the author, even if it's you ;)
 6          print('ARM by JT.')
 7
 8          # This is a definition of a network. See the next cell.
 9          self.net = net
10          # This is how many values a pixel can take.
11          self.num_vals = num_vals
12          # This is the problem dimentionality (the number of
    pixels)
13          self.D = D
14
15      # This function calculates the arm output.
16      def f(self, x):
17          # First, we apply causal convolutions.
18          h = self.net(x.unsqueeze(1))
19          # In channels, we have the number of values. Therefore,
    we change the order of dims.
20          h = h.permute(0, 2, 1)
21          # We apply softmax to calculate probabilities.
22          p = torch.softmax(h, 2)
23          return p
24
25      # The forward pass calculates the log-probability of an image
    .
26      def forward(self, x, reduction='avg'):
27          if reduction == 'avg':
```

```
28            return -(self.log_prob(x).mean())
29        elif reduction == 'sum':
30            return -(self.log_prob(x).sum())
31        else:
32            raise ValueError('reduction could be either 'avg' or
   'sum'.')
33
34    # This function calculates the log-probability (log-
   categorical).
35    # See the full code in the separate file for details.
36    def log_prob(self, x):
37        mu_d = self.f(x)
38        log_p = log_categorical(x, mu_d, num_classes=self.
   num_vals, reduction='sum', dim=-1).sum(-1)
39
40        return log_p
41
42    # This function implements sampling procedure.
43    def sample(self, batch_size):
44        # As you can notice, we first initialize a tensor with
   zeros.
45        x_new = torch.zeros((batch_size, self.D))
46
47        # Then, iteratively, we sample a value for a pixel.
48        for d in range(self.D):
49            p = self.f(x_new)
50            x_new_d = torch.multinomial(p[:, d, :], num_samples
   =1)
51            x_new[:, d] = x_new_d[:,0]
52
53        return x_new
```

Listing 2.2   Autoregressive model parameterized by causal convolutions 1D

```
1  # An example of a network. NOTICE: The first layer is A=True,
      while all the others are A=False.
2  # At this point we should know already why :)
3  M = 256
4
5  net = nn.Sequential(
6      CausalConv1d(in_channels=1, out_channels=MM, dilation=1,
      kernel_size=kernel, A=True, bias=True),
7      nn.LeakyReLU(),
8      CausalConv1d(in_channels=MM, out_channels=MM, dilation=1,
      kernel_size=kernel, A=False, bias=True),
9      nn.LeakyReLU(),
10     CausalConv1d(in_channels=MM, out_channels=MM, dilation=1,
      kernel_size=kernel, A=False, bias=True),
11     nn.LeakyReLU(),
12     CausalConv1d(in_channels=MM, out_channels=num_vals, dilation
      =1, kernel_size=kernel, A=False, bias=True))
```

Listing 2.3   An example of a network

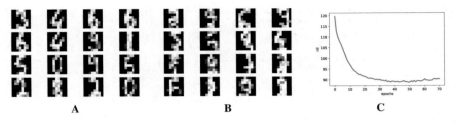

Fig. 2.4 An example of outcomes after the training: (**a**) Randomly selected real images. (**b**) Unconditional generations from the ARM. (**c**) The validation curve during training

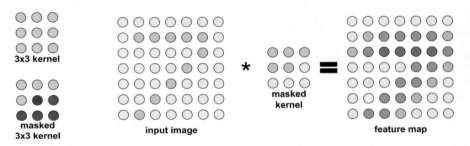

Fig. 2.5 An example of a masked 3×3 kernel (i.e., a causal 2D kernel): (*left*) A difference between a standard kernel (all weights are used; denoted by green) and a masked kernel (some weights are masked, i.e., not used; in red). For the masked kernel, we denoted the node (pixel) in the middle in violet, because it is either masked (option A) or not (option B). (*middle*) An example of an image (light orange nodes: zeros, light blue nodes: ones) and a masked kernel (option A). (*right*) The result of applying the masked kernel to the image (with padding equal to 1)

Perfect! Now we are ready to run the full code. After training our ARM, we should obtain results similar to those in Fig. 2.4.

2.4   Is It All? No!

First of all, we discussed one-dimensional causal convolutions that are typically insufficient for modeling images due to their spatial dependencies in 2D (or 3D if we consider more than 1 channel; for simplicity, we focus on a 2D case). In [10], a CausalConv2D was proposed. The idea is similar to that discussed so far, but now we need to ensure that the kernel will not look into future pixels in both the x-axis and y-axis. In Fig. 2.5, we present the difference between a standard kernel where all kernel weights are used and a masked kernel with some weights zeroed-out (or masked). Notice that in CausalConv2D we must also use option A for the first layer (i.e., we skip the pixel in the middle) and we can pick option B for the remaining layers. In Fig. 2.6, we present the same example as in Fig. 2.5 but using numeric values.

Fig. 2.6 The same example
as in Fig. 2.5 but with
numeric values

$$
\begin{bmatrix}
0 & 0 & 0 & 0 & 0 & 0 & 0 \\
0 & 1 & 1 & 1 & 1 & 1 & 0 \\
0 & 0 & 0 & 0 & 0 & 1 & 0 \\
0 & 0 & 0 & 0 & 1 & 0 & 0 \\
0 & 0 & 0 & 1 & 0 & 0 & 0 \\
0 & 0 & 1 & 0 & 0 & 0 & 0 \\
0 & 1 & 0 & 0 & 0 & 0 & 0
\end{bmatrix}
*
\begin{bmatrix}
1 & 1 & 1 \\
1 & 1 & 0 \\
0 & 0 & 0
\end{bmatrix}
=
\begin{bmatrix}
0 & 0 & 0 & 0 & 0 & 0 & 0 \\
0 & 1 & 2 & 2 & 2 & 1 & 0 \\
1 & 2 & 3 & 3 & 3 & 3 & 2 \\
0 & 0 & 0 & 0 & 2 & 2 & 1 \\
0 & 0 & 0 & 2 & 2 & 1 & 0 \\
0 & 0 & 2 & 2 & 1 & 0 & 0 \\
0 & 2 & 2 & 1 & 0 & 0 & 0
\end{bmatrix}
$$

In [12], the authors propose a further improvement on the causal convolutions. The main idea relies on creating a block that consists of vertical and horizontal convolutional layers. Moreover, they use gated non-linearity function, namely:

$$
\mathbf{h} = \tanh(\mathbf{W}\mathbf{x}) \odot \sigma(\mathbf{V}\mathbf{x}). \tag{2.16}
$$

See Figure 2 in [12] for details.

Further improvements on ARMs applied to images are presented in [13]. Therein, the authors propose to replace the categorical distribution used for modeling pixel values with the discretized logistic distribution. Moreover, they suggest to use a mixture of discretized logistic distributions to further increase flexibility of their ARMs.

The introduction of the causal convolution opened multiple opportunities for deep generative modeling and allowed obtaining state-of-the-art generations and density estimations. It is impossible to review all papers here, we just name a few interesting directions/applications that are worth remembering:

- An alternative ordering of pixels was proposed in [14]. Instead of using the ordering from left to right, a "zig–zag" pattern was proposed that allows pixels to depend on pixels previously sampled to the left and above.
- ARMs could be used as stand-alone models or they can be used in a combination with other approaches. For instance, they can be used for modeling a prior in the (Variational) Auto-Encoders [15].
- ARMs could be also used to model videos [16]. Factorization of sequential data like video is very natural, and ARMs fit this scenario perfectly.
- A possible drawback of ARMs is a lack of latent representation because all conditionals are modeled explicitly from data. To overcome this issue, [17] proposed to use a PixelCNN-based decoder in a Variational Auto-Encoder.
- An interesting and important research direction is about proposing new architectures/components of ARMs or speeding them up. As mentioned earlier, sampling from ARMs could be slow, but there are ideas to improve on that by predictive sampling [11, 18].
- Alternatively, we can replace the likelihood function with other similarity metrics, e.g., the Wasserstein distance between distributions as in quantile regression. In the context of ARMs, quantile regression was applied in [19], requiring only minor architectural changes, that resulted in improved quality scores.

- An important class of models constitute *transformers* [20]. These models use self-attention layers instead of causal convolutions.
- Multi-scale ARMs were proposed to scale high-quality images logarithmically instead of quadratically. The idea is to make local independence assumptions [21] or impose a partitioning on the spatial dimensions [22]. Even though these ideas allow lowering the memory requirements, sampling remains rather slow.

References

1. Junyoung Chung, Caglar Gulcehre, KyungHyun Cho, and Yoshua Bengio. Empirical evaluation of gated recurrent neural networks on sequence modeling. *arXiv preprint arXiv:1412.3555*, 2014.
2. Sepp Hochreiter and Jürgen Schmidhuber. Long short-term memory. *Neural computation*, 9(8):1735–1780, 1997.
3. Ilya Sutskever, James Martens, and Geoffrey E Hinton. Generating text with recurrent neural networks. In *ICML*, 2011.
4. Razvan Pascanu, Tomas Mikolov, and Yoshua Bengio. On the difficulty of training recurrent neural networks. In *International conference on machine learning*, pages 1310–1318. PMLR, 2013.
5. Martin Arjovsky, Amar Shah, and Yoshua Bengio. Unitary evolution recurrent neural networks. In *International Conference on Machine Learning*, pages 1120–1128. PMLR, 2016.
6. Ronan Collobert and Jason Weston. A unified architecture for natural language processing: Deep neural networks with multitask learning. In *Proceedings of the 25th international conference on Machine learning*, pages 160–167, 2008.
7. Nal Kalchbrenner, Edward Grefenstette, and Phil Blunsom. A convolutional neural network for modelling sentences. In *Proceedings of the 52nd Annual Meeting of the Association for Computational Linguistics*, pages 212–217. Association for Computational Linguistics, 2014.
8. Shaojie Bai, J Zico Kolter, and Vladlen Koltun. An empirical evaluation of generic convolutional and recurrent networks for sequence modeling. *arXiv preprint arXiv:1803.01271*, 2018.
9. Aaron van den Oord, Sander Dieleman, Heiga Zen, Karen Simonyan, Oriol Vinyals, Alex Graves, Nal Kalchbrenner, Andrew Senior, and Koray Kavukcuoglu. Wavenet: A generative model for raw audio. *arXiv preprint arXiv:1609.03499*, 2016.
10. Aaron Van Oord, Nal Kalchbrenner, and Koray Kavukcuoglu. Pixel recurrent neural networks. In *International Conference on Machine Learning*, pages 1747–1756. PMLR, 2016.
11. Auke Wiggers and Emiel Hoogeboom. Predictive sampling with forecasting autoregressive models. In *International Conference on Machine Learning*, pages 10260–10269. PMLR, 2020.
12. Aäron van den Oord, Nal Kalchbrenner, Oriol Vinyals, Lasse Espeholt, Alex Graves, and Koray Kavukcuoglu. Conditional image generation with pixelcnn decoders. In *Proceedings of the 30th International Conference on Neural Information Processing Systems*, pages 4797–4805, 2016.
13. Tim Salimans, Andrej Karpathy, Xi Chen, and Diederik P Kingma. Pixelcnn++: Improving the pixelcnn with discretized logistic mixture likelihood and other modifications. *arXiv preprint arXiv:1701.05517*, 2017.
14. Xi Chen, Nikhil Mishra, Mostafa Rohaninejad, and Pieter Abbeel. Pixelsnail: An improved autoregressive generative model. In *International Conference on Machine Learning*, pages 864–872. PMLR, 2018.
15. Amirhossein Habibian, Ties van Rozendaal, Jakub M Tomczak, and Taco S Cohen. Video compression with rate-distortion autoencoders. In *Proceedings of the IEEE/CVF International Conference on Computer Vision*, pages 7033–7042, 2019.

16. Nal Kalchbrenner, Aäron Oord, Karen Simonyan, Ivo Danihelka, Oriol Vinyals, Alex Graves, and Koray Kavukcuoglu. Video pixel networks. In *International Conference on Machine Learning*, pages 1771–1779. PMLR, 2017.
17. Ishaan Gulrajani, Kundan Kumar, Faruk Ahmed, Adrien Ali Taiga, Francesco Visin, David Vazquez, and Aaron Courville. PixelVAE: A latent variable model for natural images. *arXiv preprint arXiv:1611.05013*, 2016.
18. Yang Song, Chenlin Meng, Renjie Liao, and Stefano Ermon. Accelerating feedforward computation via parallel nonlinear equation solving. In *International Conference on Machine Learning*, pages 9791–9800. PMLR, 2021.
19. Georg Ostrovski, Will Dabney, and Rémi Munos. Autoregressive quantile networks for generative modeling. In *International Conference on Machine Learning*, pages 3936–3945. PMLR, 2018.
20. Ashish Vaswani, Noam Shazeer, Niki Parmar, Jakob Uszkoreit, Llion Jones, Aidan N Gomez, Łukasz Kaiser, and Illia Polosukhin. Attention is all you need. In *Advances in neural information processing systems*, pages 5998–6008, 2017.
21. Scott Reed, Aäron Oord, Nal Kalchbrenner, Sergio Gómez Colmenarejo, Ziyu Wang, Yutian Chen, Dan Belov, and Nando Freitas. Parallel multiscale autoregressive density estimation. In *International Conference on Machine Learning*, pages 2912–2921. PMLR, 2017.
22. Jacob Menick and Nal Kalchbrenner. Generating high fidelity images with subscale pixel networks and multidimensional upscaling. In *International Conference on Learning Representations*, 2018.

Chapter 3
Flow-Based Models

3.1 Flows for Continuous Random Variables

3.1.1 Introduction

So far, we have discussed a class of deep generative models that model the distribution $p(\mathbf{x})$ directly in an autoregressive manner. The main advantage of ARMs is that they can learn long-range statistics and, in a consequence, powerful density estimators. However, their drawback is that they are parameterized in an autoregressive manner, hence, sampling is rather a slow process. Moreover, they lack a latent representation, therefore, it is not obvious how to manipulate their internal data representation that makes it less appealing for tasks like compression or metric learning. In this chapter, we present a different approach to direct modeling of $p(\mathbf{x})$. However, before we start our considerations, we will discuss a simple example.

Example 3.1 Let us take a random variable $z \in \mathbb{R}$ with $\pi(z) = \mathcal{N}(z|0, 1)$. Now, we consider a new random variable after applying some linear transformation to z, namely $x = 0.75z + 1$. Now the question is the following:

What is the distribution of x, $p(x)$?

We can guess the solution by using properties of Gaussians, or dig in our memory about the **change of variables formula** to calculate this distribution, that is:

$$p(x) = \pi\left(z = f^{-1}(x)\right)\left|\frac{\partial f^{-1}(x)}{\partial x}\right|, \tag{3.1}$$

where f is an invertible function (a bijection). What does it mean? It means that the function maps one point to another, distinctive point, and we can always invert the function to obtain the original point.

© The Author(s), under exclusive license to Springer Nature Switzerland AG 2022
J. M. Tomczak, *Deep Generative Modeling*,
https://doi.org/10.1007/978-3-030-93158-2_3

Fig. 3.1 An example of a bijection where for each point in the blue set there is precisely one corresponding point in the purple set (and vice versa)

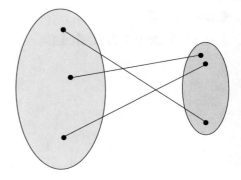

In Fig. 3.1, we have an example of a bijection. Notice that volumes of the domains do not need to be the same! Keep it in mind and think about it in the context of $\left| \frac{\partial f^{-1}(x)}{\partial x} \right|$.

Coming back to our example, we have

$$f(z) = 0.75z + 1, \tag{3.2}$$

and the inverse of f is

$$f^{-1}(x) = \frac{x - 1}{0.75}. \tag{3.3}$$

Then, the derivative of the **change of volume** is

$$\left| \frac{\partial f^{-1}(x)}{\partial x} \right| = \frac{4}{3}. \tag{3.4}$$

Putting all information so far together yields

$$p(x) = \pi \left(z = \frac{x - 1}{0.75} \right) \frac{4}{3} = \frac{1}{\sqrt{2\pi \, 0.75^2}} \exp \left\{ -(x - 1)^2 / 0.75^2 \right\}. \tag{3.5}$$

We immediately realize that we end up with the Gaussian distribution again:

$$p(x) = \mathcal{N}(x | 1, 0.75). \tag{3.6}$$

Moreover, we see that the part $\left| \frac{\partial f^{-1}(x)}{\partial x} \right|$ is responsible to **normalize** the distribution $\pi(z)$ after applying the transformation f. In other words, $\left| \frac{\partial f^{-1}(x)}{\partial x} \right|$ counteracts a possible *change of volume* caused by f. ∎

First of all, this example indicates that we can calculate a new distribution of a continuous random variable by applying a known bijective transformation f to a random variable with a known distribution, $z \sim p(z)$. The same holds for multiple variables $\mathbf{x}, \mathbf{z} \in \mathbb{R}^D$:

$$p(\mathbf{x}) = p\left(\mathbf{z} = f^{-1}(\mathbf{x})\right) \left| \frac{\partial f^{-1}(\mathbf{x})}{\partial \mathbf{x}} \right|, \tag{3.7}$$

where:

$$\left| \frac{\partial f^{-1}(\mathbf{x})}{\partial \mathbf{x}} \right| = \left| \det \mathbf{J}_{f^{-1}}(\mathbf{x}) \right| \tag{3.8}$$

is the Jacobian matrix $\mathbf{J}_{f^{-1}}$ that is defined as follows:

$$\mathbf{J}_{f^{-1}} = \begin{bmatrix} \frac{\partial f_1^{-1}}{\partial x_1} & \cdots & \frac{\partial f_1^{-1}}{\partial x_D} \\ \vdots & \ddots & \vdots \\ \frac{\partial f_D^{-1}}{\partial x_1} & \cdots & \frac{\partial f_D^{-1}}{\partial x_D} \end{bmatrix}. \tag{3.9}$$

Moreover, we can also use the **inverse function theorem** that yields

$$\left| \mathbf{J}_{f^{-1}}(\mathbf{x}) \right| = \left| \mathbf{J}_f(\mathbf{x}) \right|^{-1}. \tag{3.10}$$

Since f is invertible, we can use the inverse function theorem to rewrite (3.7) as follows:

$$p(\mathbf{x}) = p\left(\mathbf{z} = f^{-1}(\mathbf{x})\right) \left| \mathbf{J}_f(\mathbf{x}) \right|^{-1}. \tag{3.11}$$

To get some insight into the role of the Jacobian-determinant, take a look at Fig. 3.2. Here, there are three cases of invertible transformations that play around with a uniform distribution defined over a square.

In the case on top, the transformation turns a square into a rhombus without changing its volume. As a result, the Jacobian-determinant of this transformation is 1. Such transformations are called **volume-preserving**. Notice that the resulting distribution is still uniform and since there is no change of volume, it is defined over the same volume as the original one, thus, the color is the same.

In the middle, the transformation shrinks the volume, therefore, the resulting uniform distribution is "denser" (a darker color in Fig. 3.2). Additionally, the Jacobian-determinant is smaller than 1.

In the last situation, the transformation enlarges the volume, hence, the uniform distribution is defined over a larger area (a lighter color in Fig. 3.2). Since the volume is larger, the Jacobian-determinant is larger than 1.

Notice that shifting operator is volume-preserving. To see that imagine adding an arbitrary value (e.g., 5) to all points of the square. Does it change the volume? Not at all! Thus, the Jacobian-determinant equals 1.

Fig. 3.2 Three examples of invertible transformations: (*top*) a volume-preserving bijection, (*middle*) a bijection that shrinks the original area, (*bottom*) a bijection that enlarges the original area

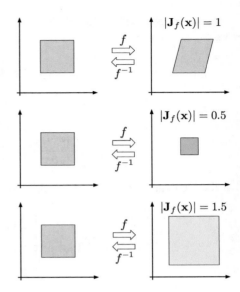

3.1.2   Change of Variables for Deep Generative Modeling

A natural question is whether we can utilize the idea of the change of variables to model a complex and high-dimensional distribution over images, audio, or other data sources. Let us consider a hierarchical model, or, equivalently, a sequence of invertible transformations, $f_k : \mathbb{R}^D \rightarrow \mathbb{R}^D$. We start with a known distribution $\pi(\mathbf{z}_0) = \mathcal{N}(\mathbf{z}_0|0, \mathbf{I})$. Then, we can sequentially apply the invertible transformations to obtain a flexible distribution [1, 2]:

$$p(\mathbf{x}) = \pi\left(\mathbf{z}_0 = f^{-1}(\mathbf{x})\right) \prod_{i=1}^{K} \left|\det \frac{\partial f_i\,(\mathbf{z}_{i-1})}{\partial \mathbf{z}_{i-1}}\right|^{-1}, \tag{3.12}$$

or by using the notation of a Jacobian for the i-th transformation:

$$p(\mathbf{x}) = \pi\left(\mathbf{z}_0 = f^{-1}(\mathbf{x})\right) \prod_{i=1}^{K} \left|\mathbf{J}_{f_i}(\mathbf{z}_{i-1})\right|^{-1}. \tag{3.13}$$

An example of transforming a unimodal base distribution like Gaussian into a multimodal distribution through invertible transformations is presented in Fig. 3.3. In principle, we should be able to get almost any arbitrarily complex distribution and revert to a *simple* one.

Let $\pi(\mathbf{z}_0)$ be $\mathcal{N}(\mathbf{z}_0|0, \mathbf{I})$. Then, the logarithm of $p(\mathbf{x})$ is the following:

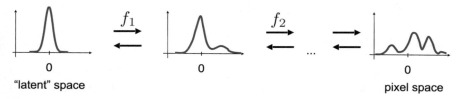

Fig. 3.3 An example of transforming a unimodal distribution (the latent space) to a multimodal distribution (the data space, e.g., the pixel space) through a series of invertible transformations f_i

$$\ln p(\mathbf{x}) = \ln \mathcal{N} \left(\mathbf{z}_0 = f^{-1}(\mathbf{x}) | 0, \mathbf{I} \right) - \sum_{i=1}^{K} \ln \left| \mathbf{J}_{f_i}(\mathbf{z}_{i-1}) \right|. \tag{3.14}$$

Interestingly, we see that the first part, namely $\ln \mathcal{N} \left(\mathbf{z}_0 = f^{-1}(\mathbf{x}) | 0, \mathbf{I} \right)$, corresponds to the *Mean Square Error* loss function between 0 and $f^{-1}(\mathbf{x})$ plus a constant. The second part, $\sum_{i=1}^{K} \ln \left| \mathbf{J}_{f_i}(\mathbf{z}_{i-1}) \right|$, as in our example, ensures that the distribution is properly normalized. However, since it penalizes the change of volume (take a look again at the example above!), we can think of it as a kind of a *regularizer* for the invertible transformations $\{f_i\}$.

Once we have laid down the foundations of the change of variables for expressing density functions, now we must face two questions:

- How to model the invertible transformations?
- What is the difficulty here?

The answer to the first question could be neural networks because they are flexible and easy-to-train. However, we cannot take **any** neural network because of two reasons. First, the transformation must be **invertible**, thus, we must pick an **invertible neural network**. Second, even if a neural network is invertible, we face a problem of calculating the second part of (3.14), i.e., $\sum_{i=1}^{K} \ln \left| \mathbf{J}_{f_i}(\mathbf{z}_{i-1}) \right|$, that is non-trivial and computationally intractable for an arbitrary sequence of invertible transformations. As a result, we seek for such neural networks that are both invertible and the logarithm of a Jacobian-determinant is (relatively) easy to calculate. The resulting model that consists of invertible transformations (neural networks) with tractable Jacobian-determinants is referred to as *normalizing flows* or *flow-based models*.

There are various possible invertible neural networks with tractable Jacobian-determinants, e.g., Planar Normalizing Flows [1], Sylvester Normalizing Flows [3], Residual Flows [4, 5], Invertible DenseNets [6]. However, here we focus on a very important class of models: **RealNVP**, *Real-valued Non-Volume Preserving* flows [7] that serve as a starting point for many other flow-based generative models (e.g., GLOW [8]).

3.1.3   Building Blocks of RealNVP

3.1.3.1   Coupling Layers

The main component of RealNVP is a *coupling layer*. The idea behind this transformation is the following. Let us consider an input to the layer that is divided into two parts: $\mathbf{x} = [\mathbf{x}_a, \mathbf{x}_b]$. The division into two parts could be done by dividing the vector $\mathbf{x}$ into $\mathbf{x}_{1:d}$ and $\mathbf{x}_{d+1:D}$ or according to a more sophisticated manner, e.g., a *checkerboard pattern* [7]. Then, the transformation is defined as follows:

$$\mathbf{y}_a = \mathbf{x}_a \tag{3.15}$$

$$\mathbf{y}_b = \exp\left(s\left(\mathbf{x}_a\right)\right) \odot \mathbf{x}_b + t\left(\mathbf{x}_a\right), \tag{3.16}$$

where $s(\cdot)$ and $t(\cdot)$ are **arbitrary neural networks** called *scaling* and *transition*, respectively.

This transformation is invertible by design, namely:

$$\mathbf{x}_b = \left(\mathbf{y}_b - t(\mathbf{y}_a)\right) \odot \exp\left(-s(\mathbf{y}_a)\right) \tag{3.17}$$

$$\mathbf{x}_a = \mathbf{y}_a. \tag{3.18}$$

Importantly, the logarithm of the Jacobian-determinant is easy to calculate, because:

$$\mathbf{J} = \begin{bmatrix} \mathbf{I}_{d \times d} & \mathbf{0}_{d \times (D-d)} \\ \frac{\partial \mathbf{y}_b}{\partial \mathbf{x}_a} & \mathrm{diag}\left(\exp\left(s\left(\mathbf{x}_a\right)\right)\right) \end{bmatrix} \tag{3.19}$$

that yields

$$\det(\mathbf{J}) = \prod_{j=1}^{D-d} \exp\left(s\left(\mathbf{x}_a\right)\right)_j = \exp\left(\sum_{j=1}^{D-d} s\left(\mathbf{x}_a\right)_j\right). \tag{3.20}$$

Eventually, coupling layers seem to be flexible and powerful transformations with tractable Jacobian-determinants! However, we process only half of the input, therefore, we must think of an appropriate additional transformation a coupling layer could be combined with.

3.1.3.2   Permutation Layers

A simple yet effective transformation that could be combined with a coupling layer is a **permutation layer**. Since permutation is *volume-preserving*, i.e., its Jacobian-

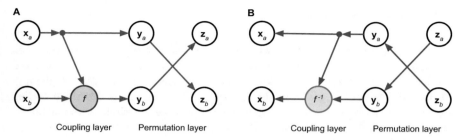

Fig. 3.4 A combination of a coupling layer and a permutation layer that transforms $[\mathbf{x}_a, \mathbf{x}_b]$ to $[\mathbf{z}_a, \mathbf{z}_b]$. (**a**) A forward pass through the block. (**b**) An inverse pass through the block

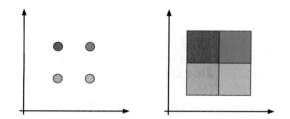

Fig. 3.5 A schematic representation of the uniform dequantization for two binary random variables: (*left*) the probability mass is assigned to points, (*right*) after the uniform dequantization, the probability mass is assigned to square areas. Colors correspond to probability values

determinant is equal to 1, we can apply it each time after the coupling layer. For instance, we can reverse the order of variables.

An example of an invertible block, i.e., a combination of a coupling layer with a permutation layer is schematically presented in Fig. 3.4.

3.1.3.3 Dequantization

As discussed so far, flow-based models assume that $\mathbf{x}$ is a vector of real-valued random variables. However, in practice, many objects are discrete. For instance, images are typically represented as integers taking values in $\{0, 1, \ldots, 255\}^D$. In [9], it has been outlined that adding a uniform noise, $\mathbf{u} \in [-0.5, 0.5]^D$, to original data, $\mathbf{y} \in \{0, 1, \ldots, 255\}^D$, allows applying density estimation to $\mathbf{x} = \mathbf{y} + \mathbf{u}$. This procedure is known as *uniform dequantization*. Recently, there were different schema of dequantization proposed, you can read more on that in [10].

An example for two binary random variables and the uniform dequantization is depicted in Fig. 3.5. After adding $\mathbf{u} \in [-0.5, 0.5]^2$ to each discrete value, we obtain a continuous space and now probabilities originally associated with volumeless points are "spread" across small square regions.

3.1.4 Flows in Action!

Let us turn math into a code! We will first discuss the log-likelihood function (i.e., the learning objective) and how mathematical formulas correspond to the code. First, it is extremely important to know what is our learning objective, i.e., the log-likelihood function. In the example, we use coupling layers as described earlier, together with permutation layers. Then, we can plug the logarithm of the Jacobian-determinant for the coupling layers (for the permutation layers it is equal to 1, so $\ln(1) = 0$) in Eq. (3.14) that yields

$$\ln p(\mathbf{x}) = \ln \mathcal{N}\left(\mathbf{z}_0 = f^{-1}(\mathbf{x})|0, \mathbf{I}\right) - \sum_{i=1}^{K}\left(\sum_{j=1}^{D-d} s_k\left(\mathbf{x}_a^k\right)_j\right), \tag{3.21}$$

where s_k is the scale network in the k-th coupling layer, and $\mathbf{x}_a^k$ denotes the input to the k-th coupling layer. Notice that exp in the log-Jacobian-determinant is cancelled by applying the logarithm.

Let us think again about the learning objective from the implementation perspective. First, we definitely need to obtain $\mathbf{z}$ by calculating $f^{-1}(\mathbf{x})$, and then we can calculate $\ln \mathcal{N}\left(\mathbf{z}_0 = f^{-1}(\mathbf{x})|0, \mathbf{I}\right)$. That is actually easy, and we get

$$\ln \mathcal{N}\left(\mathbf{z}_0 = f^{-1}(\mathbf{x})|0, \mathbf{I}\right) = -const - \frac{1}{2}\|f^{-1}(\mathbf{x})\|^2, \tag{3.22}$$

where $const = \frac{D}{2}\ln(2\pi)$ is the normalizing constant of the standard Gaussian, and $\frac{1}{2}\|f^{-1}(\mathbf{x})\|^2 = MSE(0, f^{-1}(\mathbf{x}))$.

Alright, now we should look into the second part of the objective, i.e., the log-Jacobian-determinants. As we can see, we have a sum over transformations, and for each coupling layer, we consider only the outputs of the scale nets. Hence, the only thing we must remember during implementing the coupling layers is to return not only output but also the outcome of the scale layer too.

3.1.5 Code

Now, we have all components to implement our own RealNVP! Below, there is a code with a lot of comments that should help to understand every single line of it. The full code (with auxiliary functions) that you can play with is available here: https://github.com/jmtomczak/intro_dgm.

```
1  class RealNVP(nn.Module):
2      def __init__(self, nets, nett, num_flows, prior, D=2,
         dequantization=True):
3          super(RealNVP, self).__init__()
```

```
4
5      # Well, it's always good to brag about yourself.
6      print('RealNVP by JT.')
7
8      # We need to dequantize discrete data. This attribute is
       used during training to dequantize integer data.
9      self.dequantization = dequantization
10
11     # An object of a prior (here: torch.distribution of
       multivariate normal distribution)
12     self.prior = prior
13     # A module list for translation networks
14     self.t = torch.nn.ModuleList([nett() for _ in range(
       num_flows)])
15     # A module list for scale networks
16     self.s = torch.nn.ModuleList([nets() for _ in range(
       num_flows)])
17     # The number of transformations, in our equations it is
       denoted by K.
18     self.num_flows = num_flows
19
20     # The dimensionality of the input. It is used for
       sampling.
21     self.D = D
22
23  # This is the coupling layer, the core of the RealNVP model.
24  def coupling(self, x, index, forward=True):
25      # x: input, either images (for the first transformation)
       or outputs from the previous transformation
26      # index: it determines the index of the transformation
27      # forward: whether it is a pass from x to y (forward=True
       ), or from y to x (forward=False)
28
29      # We chunk the input into two parts: x_a, x_b
30      (xa, xb) = torch.chunk(x, 2, 1)
31
32      # We calculate s(xa), but without exp!
33      s = self.s[index](xa)
34      # We calculate t(xa)
35      t = self.t[index](xa)
36
37      # Calculate either the forward pass (x -> z) or the
       inverse pass (z -> x)
38      # Note that we use the exp here!
39      if forward:
40          #yb = f^{-1}(x)
41          yb = (xb - t) * torch.exp(-s)
42      else:
43          #xb = f(y)
44          yb = torch.exp(s) * xb + t
45
46      # We return the output y = [ya, yb], but also s for
       calculating the log-Jacobian-determinant
47      return torch.cat((xa, yb), 1), s
```

```
48
49     # An implementation of the permutation layer
50     def permute(self, x):
51         # Simply flip the order.
52         return x.flip(1)
53
54     def f(self, x):
55         # This is a function that calculates the full forward
       pass through the coupling+permutation layers.
56         # We initialize the log-Jacobian-det
57         log_det_J, z = x.new_zeros(x.shape[0]), x
58         # We iterate through all layers
59         for i in range(self.num_flows):
60             # First, do coupling layer,
61             z, s = self.coupling(z, i, forward=True)
62             # then permute.
63             z = self.permute(z)
64             # To calculate the log-Jacobian-determinant of the
       sequence of transformations we sum over all of them.
65             # As a result, we can simply accumulate individual
       log-Jacobian determinants.
66             log_det_J = log_det_J - s.sum(dim=1)
67         # We return both z and the log-Jacobian-determinant,
       because we need z to feed in to the logarithm of the Norma;
68         return z, log_det_J
69
70     def f_inv(self, z):
71         # The inverse path: from z to x.
72         # We apply all transformations in the reversed order.
73         x = z
74         for i in reversed(range(self.num_flows)):
75             x = self.permute(x)
76             x, _ = self.coupling(x, i, forward=False)
77         # Since we use this function for sampling, we don't need
       to return anything else than x.
78         return x
79
80     def forward(self, x, reduction='avg'):
81         # This function is essential for PyTorch.
82         # First, we calculate the forward part: from x to z, and
       also we need the log-Jacobian-determinant.
83         z, log_det_J = self.f(x)
84         # We can use either sum or average as the output.
85         # Either way, we calculate the learning objective: self.
       prior.log_prob(z) + log_det_J.
86         # NOTE: Mind the minus sign! We need it, because, by
       default, we consider the minimization problem,
87         # but normally we look for the maximum likelihood
       estimate. Therefore, we use:
88         # max F(x) <=> min -F(x)
89         if reduction == 'sum':
90             return -(self.prior.log_prob(z) + log_det_J).sum()
91         else:
92             return -(self.prior.log_prob(z) + log_det_J).mean()
```

```
93
94      def sample(self, batchSize):
95          # First, we sample from the prior, z ~ p(z) = Normal(z
            |0,1)
96          z = self.prior.sample((batchSize, self.D))
97          z = z[:, 0, :]
98          # Second, we go from z to x.
99          x = self.f_inv(z)
100         return x.view(-1, self.D)
```

Listing 3.1   An example of an implementation of RealNVP

```
1  # The number of flows
2  num_flows = 8
3
4  # Neural networks for a single transformation (a single flow).
5  nets = lambda: nn.Sequential(nn.Linear(D//2, M), nn.LeakyReLU(),
6                               nn.Linear(M, M), nn.LeakyReLU(),
7                               nn.Linear(M, D//2), nn.Tanh())
8
9  nett = lambda: nn.Sequential(nn.Linear(D//2, M), nn.LeakyReLU(),
10                               nn.Linear(M, M), nn.LeakyReLU(),
11                               nn.Linear(M, D//2))
12
13 # For the prior, we can use the built-in PyTorch distribution.
14 prior = torch.distributions.MultivariateNormal(torch.zeros(D),
        torch.eye(D))
15
16 # Init of the RealNVP. Please note that we need to dequantize the
        data (i.e., uniform dequantization).
17 model = RealNVP(nets, nett, num_flows, prior, D=D, dequantization
        =True)
```

Listing 3.2   An example of networks

Et voila! Now we are ready to run the full code. After training our RealNVP, we should obtain results resembling those in Fig. 3.6.

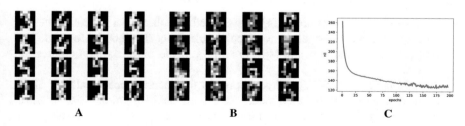

| A | B | C |

Fig. 3.6   An example of outcomes after the training: (**a**) Randomly selected real images. (**b**) Unconditional generations from the RealNVP. (**c**) The validation curve during training

3.1.6   Is It All? Really?

Yes and no. Yes in the sense it is the minimalistic example of an implementation of the RealNVP. No, because there are many improvements over the instance of the RealNVP presented here, namely:

- *Factoring-out* [7]: During the forward pass (from **x** to **z**), we can split the variables and proceed with processing only a subset of them. This could help to parameterize the base distribution by using the outputs of intermediate layers. In other words, we can obtain an autoregressive base distribution.
- *Rezero trick* [11]: Introducing additional parameters to the coupling layer, e.g., $\mathbf{y}_b = \exp(\alpha s(\mathbf{x}_a)) \odot \mathbf{x}_b + \beta t(\mathbf{x}_a)$ and α, β are initialized with 0's. This helps to ensure that the transformations act as identity maps in the beginning. It is shown in [12] that this trick helps to learn better transformations by maintaining information about the input through all layers in the beginning of the training process.
- *Masking* or *Checkerboard pattern* [7]: We can use a checkerboard pattern instead of dividing an input into two parts like $[\mathbf{x}_{1:D/2}, \mathbf{x}_{D/2+1:D}]$. This encourages learning local statistics better.
- *Squeezing* [7]: We can also play around with "squeezing" some dimensions. For instance, an image that consists of C channels, width W, and height H could be turned into $4C$ channels, width $W/2$, and height $H/2$.
- *Learnable base distributions*: instead of using a standard Gaussian base distribution, we can consider another model for that, e.g., an autoregressive model.
- *Invertible 1x1 convolution* [8]: A fixed permutation could be replaced with a (learned) invertible 1x1 convolution as in the GLOW model [8].
- *Variational dequantization* [13]: We can also pick a different dequantization scheme, e.g., variational dequantization. This allows to obtain much better scores. However, it is not for free because it leads to a lower bound to the log-likelihood function.

Moreover, there are many new fascinating research directions! I will name them here and point to papers where you can find more details:

- *Data compression with flows* [14]: Flow-based models are perfect candidates for compression since they allow to calculate the exact likelihood. Ho et al. [14] proposed a scheme that allows to use flows in the bit-back-like compression scheme.
- *Conditional flows* [15–17]: Here, we present the unconditional RealNVP. However, we can use a flow-based model for conditional distributions. For instance, we can use the conditioning as an input to the scale network and the translation network.
- *Variational inference with flows* [1, 3, 18–21]: Conditional flow-based models could be used to form a flexible family of variational posteriors. Then, the lower bound to the log-likelihood function could be tighter. We will come back to that in Chap. 4, Sect. 4.4.2.

- *Integer discrete flows* [12, 22, 23]: Another interesting direction is a version of the RealNVP for integer-valued data. We will explain this idea in Sect. 3.2.
- *Flows on manifolds* [24]: Typically, flow-based models are considered in the Euclidean space. However, they could be considered in non-Euclidean spaces, resulting in new properties of (partially) invertible transformations.
- *Flows for ABC* [25]: Approximate Bayesian Computation (ABC) assumes that the posterior over quantities of interest is intractable. One possible approach to mitigate this issue is to approximate it using flow-based models, e.g., masked autoregressive flows [26], as presented in [25].

Many other interesting information on flow-based models could be found in a fantastic review by Papamakarios et al. [27].

3.1.7 ResNet Flows and DenseNet Flows

ResNet Flows [4, 5]
In the previous sections, we have discussed flow-based models with a pre-designed architectures (i.e., blocks consisting of coupling layers and permutation layers) that allow easy calculation of the Jacobian-determinant. However, we can take a different approach and think of how we can approximate the Jacobian-determinant for an almost arbitrary architecture. And, additionally, what kind of requirements we must impose to make the architecture invertible.

In [4], the authors consider widely used residual neural networks (ResNets) and construct an invertible ResNet layer which is only constrained in Lipschitz continuity. A ResNet is defined as: $F(x) = x + g(x)$, where g is modeled by a (convolutional) neural network and F represents a ResNet layer which is in general not invertible. However, g is constructed in such a way that it satisfies the Lipschitz constant being strictly lower than 1, $\text{Lip}(g) < 1$, by using spectral normalization of [28, 29]:

$$\text{Lip}(g) < 1, \quad \text{if} \quad ||W_i||_2 < 1, \tag{3.23}$$

where $|| \cdot ||_2$ is the ℓ_2 matrix norm. Then $\text{Lip}(g) = K < 1$ and $\text{Lip}(F) < 1 + K$. Only in this specific case the Banach fixed-point theorem holds and ResNet layer F has a unique inverse. As a result, the inverse can be approximated by fixed-point iterations [4].

To estimate the log-determinant is, especially for high-dimensional spaces, computationally intractable due to expensive computations. Since ResNet blocks have a constrained Lipschitz constant, the logarithm of the Jacobian-determinant is cheaper to compute, tractable, and approximated with guaranteed convergence [4]:

$$\ln p(x) = \ln p(f(x)) + \text{tr}\left(\sum_{k=1}^{\infty} \frac{(-1)^{k+1}}{k}[J_g(x)]^k\right), \tag{3.24}$$

where $J_g(x)$ is the Jacobian of g at x that satisfies $||J_g|| < 1$. The Skilling-Hutchinson trace estimator [30, 31] is used to compute the trace at a lower cost than to fully compute the trace of the Jacobian. Residual Flows [5] use an improved method to estimate the power series at an even lower cost with an unbiased estimator based on "Russian roulette" of [32]. Intuitively, the method estimates the infinite sum of the power series by evaluating a finite amount of terms. In return, this leads to less computation of terms compared to invertible residual networks. To avoid derivative saturation, which occurs when the second derivative is zero in large regions, the LipSwish activation is proposed [4].

DenseNet Flows [6]

Since it is possible to formulate a flow for a ResNet architecture, a natural question is whether it could be accomplished for densely connected networks (DensNets) [33]. In [6], it was shown that indeed it is possible!

The main component of DenseNet flows is a DenseBlock that is defined as a function $F : \mathbb{R}^d \to \mathbb{R}^d$ with $F(x) = x + g(x)$, where g consists of dense layers $\{h_i\}_{i=1}^n$. Note that an important modification to make the model invertible is to output $x + g(x)$, whereas a standard DenseBlock would only output $g(x)$. The function g is expressed as follows:

$$g(x) = h_{n+1} \circ h_n \circ \cdots \circ h_1(x), \tag{3.25}$$

where h_{n+1} represents a 1×1 convolution to match the output size of $\mathbb{R}^d$. A layer h_i consists of two parts concatenated to each other. The upper part is a copy of the input signal. The lower part consists of the transformed input, where the transformation is a multiplication of (convolutional) weights W_i with the input signal, followed by a non-linearity ϕ having $\text{Lip}(\phi) \le 1$, such as ReLU, ELU, LipSwish, or tanh. As an example, a dense layer h_2 can be composed as follows:

$$h_1(x) = \begin{bmatrix} x \\ \phi(W_1 x) \end{bmatrix}, \; h_2(h_1(x)) = \begin{bmatrix} h_1(x) \\ \phi(W_2 h_1(x)) \end{bmatrix}. \tag{3.26}$$

The DenseNet flows [6] rely on the same techniques for approximating the Jacobian-determinant as in the ResNet flows. The main difference between the DenseNet flows and the ResNet flows lies in normalizing weights so that the Lipschitz constant of the transformation is smaller than 1 and, thus, the transformation is invertible. Formally, to satisfy $\text{Lip}(g) < 1$, we need to enforce $\text{Lip}(h_i) < 1$ for all n layers, since $\text{Lip}(g) \le \text{Lip}(h_{n+1}) \cdot \ldots \cdot \text{Lip}(h_1)$. Therefore, we first need to determine the Lipschitz constant for a dense layer h_i. We know that a function f is K-Lipschitz if for all points v and w the following holds:

$$d_Y(f(v), f(w)) \leq K d_X(v, w), \tag{3.27}$$

where we assume that the distance metrics $d_X = d_Y = d$ are chosen to be the ℓ_2-norm. Further, let two functions f_1 and f_2 be concatenated in h:

$$h_v = \begin{bmatrix} f_1(v) \\ f_2(v) \end{bmatrix}, \quad h_w = \begin{bmatrix} f_1(w) \\ f_2(w) \end{bmatrix}, \tag{3.28}$$

where function f_1 is the upper part and f_2 is the lower part. We can now find an analytical form to express a limit on K for the dense layer in the form of Eq. (3.27):

$$
\begin{aligned}
d(h_v, h_w)^2 &= d(f_1(v), f_1(w))^2 + d(f_2(v), f_2(w))^2, \\
d(h_v, h_w)^2 &\leq \left(K_1^2 + K_2^2 \right) d(v, w)^2,
\end{aligned}
\tag{3.29}
$$

where we know that the Lipschitz constant of h consist of two parts, namely $\text{Lip}(f_1) = K_1$ and $\text{Lip}(f_2) = K_2$. Therefore, the Lipschitz constant of layer h can be expressed as

$$\text{Lip}(h) = \sqrt{\left(K_1^2 + K_2^2 \right)}. \tag{3.30}$$

With spectral normalization of Eq. (3.23), we know that we can enforce (convolutional) weights W_i to be at most 1-Lipschitz. Hence, for all n dense layers we apply the spectral normalization on the lower part which locally enforces $\text{Lip}(f_2) = K_2 < 1$. Further, since we enforce each layer h_i to be at most 1-Lipschitz and we start with h_1, where $f_1(x) = x$, we know that $\text{Lip}(f_1) = 1$. Therefore, the Lipschitz constant of an entire layer can be at most $\text{Lip}(h) = \sqrt{1^2 + 1^2} = \sqrt{2}$, thus dividing by this limit enforces each layer to be at most 1-Lipschitz. To read more about DenseNet flows and further improvements, please see the original paper [6].

3.2   Flows for Discrete Random Variables

3.2.1   Introduction

While discussing flow-based models in the previous section, we presented them as *density estimators*, namely models that represent stochastic dependencies among continuous random variables. We introduced the *change of variables* formula that helps to express a random variable by transforming it using invertible maps (bijections) f to a random variable with a known probability density function. Formally, it is defined as follows:

$$p(\mathbf{x}) = p\left(\mathbf{v} = f^{-1}(\mathbf{x}) \right) \left| \mathbf{J}_f(x) \right|^{-1}, \tag{3.31}$$

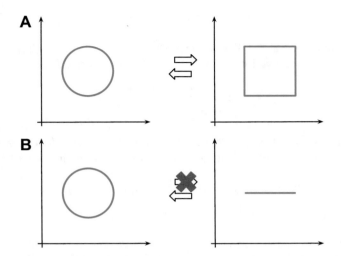

Fig. 3.7 Examples of: (**a**) homeomorphic spaces, and (**b**) non-homeomorphic spaces. The red cross indicates it is impossible to invert the transformation

where $\mathbf{J}_f(x)$ is the Jacobian of f at $\mathbf{x}$.

However, there are potential issues with such an approach. First of all, in many problems (e.g., image processing) the considered random variables (objects) are discrete. For instance, images typically take values in $\{0, 1, \ldots, 255\} \subset \mathbb{Z}$. In order to apply flows, we must apply *dequantization* [10] that results in a lower bound to the original probability distribution.

A continuous space possesses various potential pitfalls. One of them is that if we a transformation is a bijection (as in flows), not all continuous deformations are possible. It is tightly connected with *topology* and, more precisely, homeomorphisms, i.e., a continuous function between topological spaces that has a continuous inverse function, and diffeomorphisms, i.e., invertible functions that map one differentiable manifold to another such that both the function and its inverse are smooth. It is not crucial to know topology, but a curious reader may take a detour and read on that, it is definitely a fascinating field and I wish to know more about it! Anyway, let us consider three examples.

Imagine we want to transform a square into a circle (Fig. 3.7a). It is possible to find a homeomorphism (i.e., a bijection) that turns the square into the circle and back. Imagine you have a hammer and an iron square. If you start hitting the square infinitely many times, you can get an iron circle. Then, you can do it "backward" to get the square back. I know, it is unrealistic but hey, we are talking about math here!

However, if we consider a line segment and a circle (Fig. 3.7b), the situation is a bit more complicated. It is possible to transform the line segment into a circle, but not the other way around. Why? Because while transforming the circle to the line segment, it is unclear which point of the circle corresponds to the beginning (or the end) of the line segment. That is why we cannot invert the transformation!

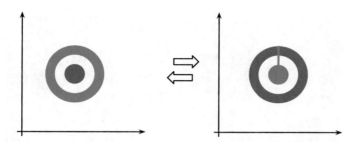

Fig. 3.8 An example of "replacing" a ring (in blue) with a ball (in magenta)

Another example that I really like, and which is closer to the potential issues of continuous flows, is transforming a ring into a ball as in Fig. 3.8. The goal is to replace the blue ring with the magenta ball. In order to make the transformation bijective, while transforming the blue ring in place of the magenta ball, we must ensure that the new magenta "ring" is in fact "broken" so that the new blue "ball" can get inside! Again, why? If the magenta ring is not broken, then we cannot say how the blue ball got inside that destroys bijectivity! In the language of topology, it is impossible because the two spaces are non-homeomorphic.

Alright, but how this affects the flow-based models? I hope that some of you asked this question, or maybe even imagined possible cases where this might hinder learning flows. In general, I would say it is fine, and we should not look for faults where there is none or almost none. However, if you work with flows that require dequantization, then you can spot cases like the one in Fig. 3.9. In this simple example, we have two discrete random variables that after uniform dequantization have two regions with equal probability mass, and the remaining two regions with zero probability mass [10]. After training a flow-based model, we have a density estimator that assigns non-zero probability mass where the true distribution has zero density! Moreover, the transformation in the flow must be a bijection, therefore, there is a continuity between the two squares (see Fig. 3.9, right). Where did we see that? Yes, in Fig. 3.8! We must know how to invert the transformation, thus, there must be a "trace" of how the probability mass moves between the regions.

Again, we can ask ourselves if it is bad. Well, I would say not really, but if we think of a case with more random variables, and there is always some little error here and there, this causes a *probability mass leakage* that could result in a far-from-perfect model. And, overall, the model could err in proper probability assignment.

3.2.2 Flows in $\mathbb{R}$ or Maybe Rather in $\mathbb{Z}$?

Before we consider any specific cases and discuss discrete flows, first we need to answer whether there is a change of variables formula for discrete random variables. The answer, fortunately, is yes! Let us consider $\mathbf{x} \in \mathcal{X}^D$ where $\mathcal{X}$ is a discrete space,

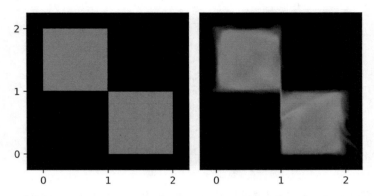

Fig. 3.9 An example of uniformly dequantized discrete random variables (*left*) and a flow-based model (*right*). Notice that in these examples, the true distribution assigns equal probability mass to the two regions in orange, and zero probability mass to the remaining two regions (in black). However, the flow-based model assigns probability mass outside the original non-zero probability regions

e.g., $\mathcal{X} = \{0, 1\}$ or $\mathcal{X} = \mathbb{Z}$. Then the change of variables takes the following form:

$$p(\mathbf{x}) = \pi \left(\mathbf{z}_0 = f^{-1}(\mathbf{x}) \right), \qquad (3.32)$$

where f is an invertible transformation and $\pi(\cdot)$ is a base distribution. Immediately we can spot a "missing" Jacobian-determinant. This is correct! Why? Because now we live in the discrete world where the probability mass is assigned to points that are "shapeless" and the bijection cannot change the volume. Thus, the Jacobian-determinant is always equal to 1! That seems to be good news, isn't it? We can take any bijective transformations and we do not need to bother about the Jacobian. That is obviously true, however, we need to remember that the output of the transformation must be still discrete, i.e., $z \in \mathcal{X}^D$. As a result, we cannot use any arbitrary invertible neural network. We will discuss it in a minute, however, before we do that, it is worth discussing the expressivity of discrete flows.

Let us assume that we have an invertible transformation $f : \mathcal{X}^D \rightarrow \mathcal{X}^D$. Moreover, we have $\mathcal{X} = \{0, 1\}$. As noted by Papamakarios et al. [27], a discrete flow can only permute probability masses. Since there is no Jacobian (or, rather, the Jacobian-determinant is equal to 1), there is no chance to decrease or increase the probability for specific values. We depict it in Fig. 3.10. You can easily imagine the situation as the space is the Rubik's cube and your hands are the flows. If you record your moves, you can always play the video backward, thus, it is invertible. However, you can only shuffle the colors around! As a result, we do not gain anything by

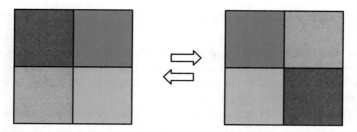

Fig. 3.10 An example of a discrete flow for two binary random variables. Colors represent various probabilities (i.e., the sum of all squares is 1)

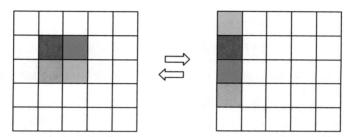

Fig. 3.11 An example of a discrete flow for two binary random variables but in the extended space. Colors represent various probabilities (i.e., the sum of all squares is 1)

applying the discrete flow, and learning the discrete flow is equivalent to learning the base distribution π.[1] So we are back to square one.

However, as pointed out by van den Berg et al. [12], the situation looks differently if we consider an extended space (or infinite space like $\mathbb{Z}$). The discrete flow can still only shuffle the probabilities, but now it can re-organize them in such a way that the probabilities can be factorized! In other words, it can help the base distribution to be a product of marginals, $\pi(\mathbf{z}) = \prod_{d=1}^{D} \pi_d(z_d|\theta_d)$, and the dependencies among variables are now encoded in the invertible transformations. An example of this case is presented in Fig. 3.11. We refer to [12] for a more thorough discussion with an appropriate lemma.

This is amazing information! It means that building a flow-based model in the discrete space makes sense. Now we can think of how to build an invertible neural network in discrete spaces and we have it!

---

[1] Well, this is not entirely true, we can still learn some correlations but it is definitely highly limited.

3.2.3 Integer Discrete Flows

We know now that it makes sense to work with discrete flows and that they are flexible as long as we use extended spaces or infinite spaces like $\mathbb{Z}$. However, the question is how to formulate an invertible transformation (or rather: an invertible neural network) that will output discrete values.

Hoogeboom et al. [22] proposed to focus on integers since they can be seen as discretized continuous values. As such, we consider coupling layers [7] and modify them accordingly. Let us remind ourselves the definition of bipartite coupling layers for $\mathbf{x} \in \mathbb{R}^D$:

$$\mathbf{y}_a = \mathbf{x}_a \tag{3.33}$$

$$\mathbf{y}_b = \exp\left(s\left(\mathbf{x}_a\right)\right) \odot \mathbf{x}_b + t\left(\mathbf{x}_a\right), \tag{3.34}$$

where $s(\cdot)$ and $t(\cdot)$ are arbitrary neural networks called *scaling* and *transition*, respectively.

Considering integer-valued variables, $\mathbf{x} \in \mathbb{Z}^D$, requires modifying this transformation. First, using scaling might be troublesome because multiplying by integers is still possible, but when we invert the transformation, we divide by integers, and dividing an integer by an integer does not necessarily result in an integer. Therefore, we must remove scaling just in case. Second, we use an arbitrary neural network for the transition. However, this network must return integers! Hoogeboom et al. [22] utilize a relatively simple trick, namely they say that we can round the output of $t(\cdot)$ to the closest integer. As a result, we add (in the forward) or subtract (in the inverse) integers from integers that is perfectly fine. (the outcome is still integer-valued.) Eventually, we get the following bipartite coupling layer:

$$\mathbf{y}_a = \mathbf{x}_a \tag{3.35}$$

$$\mathbf{y}_b = \mathbf{x}_b + \lfloor t\left(\mathbf{x}_a\right) \rceil, \tag{3.36}$$

where $\lfloor \cdot \rceil$ is the rounding operator. An inquisitive reader could ask at this point whether the rounding operator still allows using the backpropagation algorithm. In other words, whether the rounding operator is differentiable. The answer is **no**, but [22] showed that using the straight-through estimator (STE) of a gradient is sufficient. As a side note, the STE in this case uses the rounding in the forward pass of the network, $\lfloor t\left(\mathbf{x}_a\right) \rceil$, but it utilizes $t\left(\mathbf{x}_a\right)$ in the backward pass (to calculate gradients). van den Berg et al. [12] further indicated that indeed the STE works well and the bias does not hinder training too much. The implementation of the rounding operator using the STE is presented below.

```
1 # We need to turn torch.round (i.e., the rounding operator) into
     a differentiable function. For this purpose, we use the
     rounding in the forward pass, but the original input for the
     backward pass. This is nothing else than the STE.
2 class RoundStraightThrough(torch.autograd.Function):
```

```
3
4    def __init__(self):
5        super().__init__()
6
7    @staticmethod
8    def forward(ctx, input):
9        rounded = torch.round(input, out=None)
10       return rounded
11
12   @staticmethod
13   def backward(ctx, grad_output):
14       grad_input = grad_output.clone()
15       return grad_input
```

Listing 3.3 An implementation of the rounding operator using the STE

In [23] it has been shown how to generalize invertible transformations like bipartite coupling layers, among others, namely ($X_{i:j}$ denotes a subset of X corresponding to variables from the i-th dimension to the j-th dimension, $\mathbf{x}_{i:j}$, we assume that $X_{1:0} = \emptyset$ and $X_{n+1:n} = \emptyset$):

Proposition 3.1 ([23]) *Let us take* $\mathbf{x}, \mathbf{y} \in X$. *If binary transformations* $\circ$ *and* $\triangleright$ *have inverses* $\bullet$ *and* $\blacktriangleleft$, *respectively, and* $g_2, \ldots, g_D$ *and* $f_1, \ldots, f_D$ *are arbitrary functions, where* $g_i : X_{1:i-1} \to X_i$, $f_i : X_{1:i-1} \times X_{i+1:n} \to X_i$, *then the following transformation from* $\mathbf{x}$ *to* $\mathbf{y}$:

$$y_1 = x_1 \circ f_1(\emptyset, \mathbf{x}_{2:D})$$

$$y_2 = (g_2(y_1) \triangleright x_2) \circ f_2(y_1, \mathbf{x}_{3:D})$$

$$\ldots$$

$$y_d = (g_d(\mathbf{y}_{1:d-1}) \triangleright x_d) \circ f_d(\mathbf{y}_{1:d-1}, \mathbf{x}_{d+1:D})$$

$$\ldots$$

$$y_D = (g_D(\mathbf{y}_{1:D-1}) \triangleright x_D) \circ f_D(\mathbf{y}_{1:D-1}, \emptyset)$$

is invertible.

Proof In order to inverse $\mathbf{y}$ to $\mathbf{x}$ we start from the last element to obtain the following:

$$x_D = g_D(\mathbf{y}_{1:D-1}) \blacktriangleleft (y_D \bullet f_D(\mathbf{y}_{1:D-1}, \emptyset)).$$

Then, we can proceed with next expressions in the decreasing order (i.e., from $D-1$ to 1) to eventually obtain

$$x_{D-1} = g_{D-1}(\mathbf{y}_{1:D-2}) \blacktriangleleft (y_{D-1} \bullet f_{D-1}(\mathbf{y}_{1:D-2}, x_D))$$

$$\ldots$$

$$x_d = g_d(\mathbf{y}_{1:d-1}) \blacktriangleleft (y_d \bullet f_d(\mathbf{y}_{1:d-1}, \mathbf{x}_{d+1:D}))$$

$$\dots$$

$$x_2 = g_2(y_1) \blacktriangleleft (y_2 \bullet f_2(y_1, \mathbf{x}_{3:D}))$$

$$x_1 = y_1 \bullet f_1(\emptyset, \mathbf{x}_{2:D}).$$

□

For instance, we can divide $\mathbf{x}$ into four parts, $\mathbf{x} = [\mathbf{x}_a, \mathbf{x}_b, \mathbf{x}_c, \mathbf{x}_d]$, and the following transformation (a quadripartite coupling layer) is invertible [23]:

$$\mathbf{y}_a = \mathbf{x}_a + \lfloor t(\mathbf{x}_b, \mathbf{x}_c, \mathbf{x}_d)\rceil \tag{3.37}$$

$$\mathbf{y}_b = \mathbf{x}_b + \lfloor t(\mathbf{y}_a, \mathbf{x}_c, \mathbf{x}_d)\rceil \tag{3.38}$$

$$\mathbf{y}_c = \mathbf{x}_c + \lfloor t(\mathbf{y}_a, \mathbf{y}_b, \mathbf{x}_d)\rceil \tag{3.39}$$

$$\mathbf{y}_d = \mathbf{x}_d + \lfloor t(\mathbf{y}_a, \mathbf{y}_b, \mathbf{y}_c)\rceil. \tag{3.40}$$

This new invertible transformation could be seen as a kind of autoregressive processing since $\mathbf{y}_a$ is used to calculate $\mathbf{y}_b$, then both $\mathbf{y}_a$ and $\mathbf{y}_b$ are used for obtaining $\mathbf{y}_c$ and so on. As a result, we get a more powerful transformation than the bipartite coupling layer.

If we stick to a coupling layer, we need to remember to use a permutation layer to reverse the order of variables. Otherwise, some inputs would be only partially processed. This holds true for any coupling layer, either they are used for continuous flows or integer-valued flows.

The last component we need to think of is the base distribution. Similarly to flow-based models, we can use various tricks to boost the performance of the model. For instance, we can consider squeezing, factoring-out, and a mixture model for the base distribution [22]. However, in this section, we try to keep the model as simple as possible, therefore, we use the product of marginals as the base distribution. For images represented as integers, we use the following:

$$\pi(\mathbf{z}) = \prod_{d=1}^{D} \pi_d(z_d) \tag{3.41}$$

$$= \prod_{d=1}^{D} \mathrm{DL}(z_d | \mu_d, \nu_d), \tag{3.42}$$

where $\pi_d(z_d) = \mathrm{DL}(z_d | \mu_d, \nu_d)$ is the discretized logistic distribution that is defined as a difference of CDFs of the logistic distribution as follows [34]:

$$\pi(z) = \mathrm{sigm}((z + 0.5 - \mu)/\nu) - \mathrm{sigm}((z - 0.5 - \mu)/\nu), \tag{3.43}$$

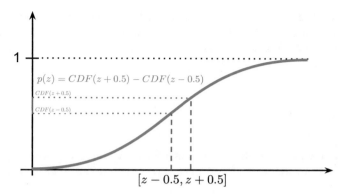

Fig. 3.12 An example of the discretized logistic distribution with $\mu = 0$ and $v = 1$. The magenta area on the y-axis corresponds to the probability mass of a bin of size 1

where $\mu \in \mathbb{R}$ and $v > 0$ denote the mean and the scale, respectively, sigm($\cdot$) is the sigmoid function. Notice that this is equivalent to calculating the probability of z falling into a bin of length 1, therefore, we add 0.5 in the first CDF and subtract 0.5 from the second CDF. An example of the discretized distribution is presented in Fig. 3.12 and the implementation follows. Interestingly, we can use this distribution to replace the Categorical distribution in Chap. 2, as it was done in [18]. We can even use a mixture of discretized logistic distribution to further improve the final performance [22, 35].

```
1 # This function implements the log of the discretized logistic
     distribution.
2 def log_integer_probability(x, mean, logscale):
3     scale = torch.exp(logscale)
4
5     logp = log_min_exp(
6         F.logsigmoid((x + 0.5 - mean) / scale),
7         F.logsigmoid((x - 0.5 - mean) / scale))
8
9     return logp
```

Listing 3.4 The logarithm of the discretized logistic distribution [34]

Eventually, our log-likelihood function takes the following form:

$$\ln p(\mathbf{x}) = \sum_{d=1}^{D} \ln \mathrm{DL}(z_d = f^{-1}(\mathbf{x})|\mu_d, v_d) \tag{3.44}$$

$$= \sum_{d=1}^{D} \ln \left(\mathrm{sigm}\left((z_d + 0.5 - \mu_d)/v_d\right) - \mathrm{sigm}\left((z_d - 0.5 - \mu_d)/v_d\right) \right), \tag{3.45}$$

where we make all μ_d and ν_d learnable parameters. Notice that ν_d must be positive (strictly larger than 0), therefore, in the implementation, we will consider the logarithm of the scale because taking exp of the log-scale ensures having strictly positive values.

3.2.4   Code

Now, we have all components to implement our own Integer Discrete Flow (IDF)! Below, there is a code with a lot of comments that should help to understand every single line of it.

```
1  # That's the class of the Integer Discrete Flows (IDFs).
2  # There are two options implemented:
3  # Option 1: The bipartite coupling layers as in (Hoogeboom et al
       ., 2019).
4  # Option 2: A new coupling layer with 4 parts as in (Tomczak,
       2021).
5  # We implement the second option explicitly, without any loop, so
        that it is very clear how it works.
6  class IDF(nn.Module):
7      def __init__(self, netts, num_flows, D=2):
8          super(IDF, self).__init__()
9
10         print('IDF by JT.')
11
12         # Option 1:
13         if len(netts) == 1:
14             self.t = torch.nn.ModuleList([netts[0]() for _ in
       range(num_flows)])
15             self.idf_git = 1
16
17         # Option 2:
18         elif len(netts) == 4:
19             self.t_a = torch.nn.ModuleList([netts[0]() for _ in
       range(num_flows)])
20             self.t_b = torch.nn.ModuleList([netts[1]() for _ in
       range(num_flows)])
21             self.t_c = torch.nn.ModuleList([netts[2]() for _ in
       range(num_flows)])
22             self.t_d = torch.nn.ModuleList([netts[3]() for _ in
       range(num_flows)])
23             self.idf_git = 4
24
25         else:
26             raise ValueError('You can provide either 1 or 4
       translation nets.')
27
28         # The number of flows (i.e., invertible transformations).
29         self.num_flows = num_flows
30
```

```
31          # The rounding operator
32          self.round = RoundStraightThrough.apply
33
34          # Initialization of the parameters of the base
       distribution.
35          # Notice they are parameters, so they are trained
       alongside the weights of neural networks.
36          self.mean = nn.Parameter(torch.zeros(1, D)) #mean
37          self.logscale = nn.Parameter(torch.ones(1, D)) #log-scale
38
39          # The dimensionality of the problem.
40          self.D = D
41
42      # The coupling layer.
43      def coupling(self, x, index, forward=True):
44
45          # Option 1:
46          if self.idf_git == 1:
47              (xa, xb) = torch.chunk(x, 2, 1)
48
49              if forward:
50                  yb = xb + self.round(self.t[index](xa))
51              else:
52                  yb = xb - self.round(self.t[index](xa))
53
54              return torch.cat((xa, yb), 1)
55
56          # Option 2:
57          elif self.idf_git == 4:
58              (xa, xb, xc, xd) = torch.chunk(x, 4, 1)
59
60              if forward:
61                  ya = xa + self.round(self.t_a[index](torch.cat((
       xb, xc, xd), 1)))
62                  yb = xb + self.round(self.t_b[index](torch.cat((
       ya, xc, xd), 1)))
63                  yc = xc + self.round(self.t_c[index](torch.cat((
       ya, yb, xd), 1)))
64                  yd = xd + self.round(self.t_d[index](torch.cat((
       ya, yb, yc), 1)))
65              else:
66                  yd = xd - self.round(self.t_d[index](torch.cat((
       xa, xb, xc), 1)))
67                  yc = xc - self.round(self.t_c[index](torch.cat((
       xa, xb, yd), 1)))
68                  yb = xb - self.round(self.t_b[index](torch.cat((
       xa, yc, yd), 1)))
69                  ya = xa - self.round(self.t_a[index](torch.cat((
       yb, yc, yd), 1)))
70
71              return torch.cat((ya, yb, yc, yd), 1)
72
73      # Similarly to RealNVP, we have also the permute layer.
74      def permute(self, x):
```

```
75        return x.flip(1)
76
77    # The main function of the IDF: forward pass from x to z.
78    def f(self, x):
79        z = x
80        for i in range(self.num_flows):
81            z = self.coupling(z, i, forward=True)
82            z = self.permute(z)
83
84        return z
85
86    # The function for inverting z to x.
87    def f_inv(self, z):
88        x = z
89        for i in reversed(range(self.num_flows)):
90            x = self.permute(x)
91            x = self.coupling(x, i, forward=False)
92
93        return x
94
95    # The PyTorch forward function. It returns the log-
      probability.
96    def forward(self, x, reduction='avg'):
97        z = self.f(x)
98        if reduction == 'sum':
99            return -self.log_prior(z).sum()
100        else:
101            return -self.log_prior(z).mean()
102
103    # The function for sampling:
104    # First we sample from the base distribution.
105    # Second, we invert z.
106    def sample(self, batchSize, intMax=100):
107        # sample z:
108        z = self.prior_sample(batchSize=batchSize, D=self.D,
      intMax=intMax)
109        # x = f^-1(z)
110        x = self.f_inv(z)
111        return x.view(batchSize, 1, self.D)
112
113    # The function for calculating the logarithm of the base
      distribution.
114    def log_prior(self, x):
115        log_p = log_integer_probability(x, self.mean, self.
      logscale)
116        return log_p.sum(1)
117
118    # A function for sampling integers from the base distribution
      .
119    def prior_sample(self, batchSize, D=2):
120        # Sample from logistic
121        y = torch.rand(batchSize, self.D)
122        # Here we use a property of the logistic distribution:
```

```
123        # In order to sample from a logistic distribution, first
        sample y ~ Uniform[0,1].
124        # Then, calculate log(y / (1.-y)), scale is with the
        scale, and add the mean.
125        x = torch.exp(self.logscale) * torch.log(y / (1. - y)) +
        self.mean
126        # And then round it to an integer.
127        return torch.round(x)
```

Listing 3.5 An example of networks

Below, we provide examples of neural networks that could be used to run the IDFs.

```
1  # The number of invertible transformations
2  num_flows = 8
3
4  # This variable defines whether we use:
5  #    Option 1: 1 - the classic coupling layer proposed in (
       Hogeboom et al., 2019)
6  #    Option 2: 4 - the general invertible transformation in (
       Tomczak, 2021) with 4 partitions
7  idf_git = 1
8
9  if idf_git == 1:
10     nett = lambda: nn.Sequential(
11                     nn.Linear(D//2, M), nn.LeakyReLU(),
12                     nn.Linear(M, M), nn.LeakyReLU(),
13                     nn.Linear(M, D//2))
14     netts = [nett]
15
16  elif idf_git == 4:
17     nett_a = lambda: nn.Sequential(
18                     nn.Linear(3 * (D//4), M), nn.LeakyReLU(),
19                     nn.Linear(M, M), nn.LeakyReLU(),
20                     nn.Linear(M, D//4))
21
22     nett_b = lambda: nn.Sequential(
23                     nn.Linear(3 * (D//4), M), nn.LeakyReLU(),
24                     nn.Linear(M, M), nn.LeakyReLU(),
25                     nn.Linear(M, D//4))
26
27     nett_c = lambda: nn.Sequential(
28                     nn.Linear(3 * (D//4), M), nn.LeakyReLU(),
29                     nn.Linear(M, M), nn.LeakyReLU(),
30                     nn.Linear(M, D//4))
31
32     nett_d = lambda: nn.Sequential(
33                     nn.Linear(3 * (D//4), M), nn.LeakyReLU(),
34                     nn.Linear(M, M), nn.LeakyReLU(),
35                     nn.Linear(M, D//4))
36
37     netts = [nett_a, nett_b, nett_c, nett_d]
38
```

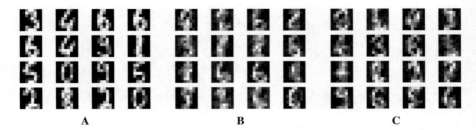

A   B   C

Fig. 3.13 An example of outcomes after the training: (**a**) Randomly selected real images. (**b**) Unconditional generations from the IDF with bipartite coupling layers. (**c**) Unconditional generations from the IDF with quadripartite coupling layers

```
39  # Init IDF
40  model = IDF(netts, num_flows, D=D)
41  # Print the summary (like in Keras)
42  print(summary(model, torch.zeros(1, 64), show_input=False,
            show_hierarchical=False))
```

Listing 3.6 An example of networks

And we are done, this is all we need to have! After running the code (take a look at: https://github.com/jmtomczak/intro_dgm) and training the IDFs, we should obtain results similar to those in Fig. 3.13.

3.2.5 What's Next?

Similarly to our example of RealNVP, here we present rather a simplified implementation of IDFs. We can use many of the tricks presented in the section on RealNVP (see Sect. 3.1.6). On recent developments on IDFs, please see also [12].

Integer discrete flows have a great potential in data compression. Since IDFs learn the distribution $p(\mathbf{x})$ directly on the integer-valued objects, they are excellent candidates for lossless compression. As presented in [22], they are competitive with other codecs for lossless compression of images.

The paper by van den Berg et al. [12] further shows that the potential bias following from the STE of the gradients is not as dramatic as originally thought [22], and they can learn flexible distributions. This result suggests that IDFs require special attention, especially for real-life applications like data compression.

It seems that the next step would be to think of more powerful transformations for discrete variables, e.g., see [23], and developing powerful architectures. Another interesting direction is utilizing alternative learning algorithms in which gradients could be better estimated [36], or even replaced [37].

References

1. Danilo Rezende and Shakir Mohamed. Variational inference with normalizing flows. In *International Conference on Machine Learning*, pages 1530–1538. PMLR, 2015.
2. Oren Rippel and Ryan Prescott Adams. High-dimensional probability estimation with deep density models. *arXiv preprint arXiv:1302.5125*, 2013.
3. Rianne Van Den Berg, Leonard Hasenclever, Jakub M Tomczak, and Max Welling. Sylvester normalizing flows for variational inference. In *34th Conference on Uncertainty in Artificial Intelligence 2018, UAI 2018*, pages 393–402. Association For Uncertainty in Artificial Intelligence (AUAI), 2018.
4. Jens Behrmann, Will Grathwohl, Ricky TQ Chen, David Duvenaud, and Jörn-Henrik Jacobsen. Invertible residual networks. In *International Conference on Machine Learning*, pages 573–582. PMLR, 2019.
5. Ricky TQ Chen, Jens Behrmann, David Duvenaud, and Jörn-Henrik Jacobsen. Residual flows for invertible generative modeling. *arXiv preprint arXiv:1906.02735*, 2019.
6. Yura Perugachi-Diaz, Jakub M Tomczak, and Sandjai Bhulai. Invertible DenseNets with concatenated LipSwish. *Advances in Neural Information Processing Systems*, 2021.
7. Laurent Dinh, Jascha Sohl-Dickstein, and Samy Bengio. Density estimation using Real NVP. *arXiv preprint arXiv:1605.08803*, 2016.
8. Diederik P Kingma and Prafulla Dhariwal. Glow: generative flow with invertible 1×1 convolutions. In *Proceedings of the 32nd International Conference on Neural Information Processing Systems*, pages 10236–10245, 2018.
9. Lucas Theis, Aäron van den Oord, and Matthias Bethge. A note on the evaluation of generative models. *arXiv preprint arXiv:1511.01844*, 2015.
10. Emiel Hoogeboom, Taco S Cohen, and Jakub M Tomczak. Learning discrete distributions by dequantization. *arXiv preprint arXiv:2001.11235*, 2020.
11. Thomas Bachlechner, Bodhisattwa Prasad Majumder, Huanru Henry Mao, Garrison W Cottrell, and Julian McAuley. Rezero is all you need: Fast convergence at large depth. *arXiv preprint arXiv:2003.04887*, 2020.
12. Rianne van den Berg, Alexey A Gritsenko, Mostafa Dehghani, Casper Kaae Sønderby, and Tim Salimans. Idf++: Analyzing and improving integer discrete flows for lossless compression. *arXiv e-prints*, pages arXiv–2006, 2020.
13. Jonathan Ho, Xi Chen, Aravind Srinivas, Yan Duan, and Pieter Abbeel. Flow++: Improving flow-based generative models with variational dequantization and architecture design. In *International Conference on Machine Learning*, pages 2722–2730. PMLR, 2019.
14. Jonathan Ho, Evan Lohn, and Pieter Abbeel. Compression with flows via local bits-back coding. *arXiv preprint arXiv:1905.08500*, 2019.
15. Michał Stypułkowski, Kacper Kania, Maciej Zamorski, Maciej Zięba, Tomasz Trzciński, and Jan Chorowski. Representing point clouds with generative conditional invertible flow networks. *arXiv preprint arXiv:2010.11087*, 2020.
16. Christina Winkler, Daniel Worrall, Emiel Hoogeboom, and Max Welling. Learning likelihoods with conditional normalizing flows. *arXiv preprint arXiv:1912.00042*, 2019.
17. Valentin Wolf, Andreas Lugmayr, Martin Danelljan, Luc Van Gool, and Radu Timofte. Deflow: Learning complex image degradations from unpaired data with conditional flows. *arXiv preprint arXiv:2101.05796*, 2021.
18. Durk P Kingma, Tim Salimans, Rafal Jozefowicz, Xi Chen, Ilya Sutskever, and Max Welling. Improved variational inference with inverse autoregressive flow. *Advances in Neural Information Processing Systems*, 29:4743–4751, 2016.
19. Emiel Hoogeboom, Victor Garcia Satorras, Jakub M Tomczak, and Max Welling. The convolution exponential and generalized Sylvester flows. *arXiv preprint arXiv:2006.01910*, 2020.
20. Jakub M Tomczak and Max Welling. Improving variational auto-encoders using householder flow. *arXiv preprint arXiv:1611.09630*, 2016.

21. Jakub M Tomczak and Max Welling. Improving variational auto-encoders using convex combination linear inverse autoregressive flow. *arXiv preprint arXiv:1706.02326*, 2017.
22. Emiel Hoogeboom, Jorn WT Peters, Rianne van den Berg, and Max Welling. Integer discrete flows and lossless compression. *arXiv preprint arXiv:1905.07376*, 2019.
23. Jakub M Tomczak. General invertible transformations for flow-based generative modeling. *INNF+*, 2021.
24. Johann Brehmer and Kyle Cranmer. Flows for simultaneous manifold learning and density estimation. *arXiv preprint arXiv:2003.13913*, 2020.
25. George Papamakarios, David Sterratt, and Iain Murray. Sequential neural likelihood: Fast likelihood-free inference with autoregressive flows. In *The 22nd International Conference on Artificial Intelligence and Statistics*, pages 837–848. PMLR, 2019.
26. George Papamakarios, Theo Pavlakou, and Iain Murray. Masked autoregressive flow for density estimation. *arXiv preprint arXiv:1705.07057*, 2017.
27. George Papamakarios, Eric Nalisnick, Danilo Jimenez Rezende, Shakir Mohamed, and Balaji Lakshminarayanan. Normalizing flows for probabilistic modeling and inference. *arXiv preprint arXiv:1912.02762*, 2019.
28. Henry Gouk, Eibe Frank, Bernhard Pfahringer, and Michael Cree. Regularisation of neural networks by enforcing Lipschitz continuity. *arXiv preprint arXiv:1804.04368*, 2018.
29. Takeru Miyato, Toshiki Kataoka, Masanori Koyama, and Yuichi Yoshida. Spectral normalization for generative adversarial networks. *arXiv preprint arXiv:1802.05957*, 2018.
30. John Skilling. The eigenvalues of mega-dimensional matrices. In *Maximum Entropy and Bayesian Methods*, pages 455–466. Springer, 1989.
31. Michael F Hutchinson. A stochastic estimator of the trace of the influence matrix for laplacian smoothing splines. *Communications in Statistics-Simulation and Computation*, 19(2):433–450, 1990.
32. Herman Kahn. Use of different Monte Carlo sampling techniques. *Proceedings of Symposium on Monte Carlo Methods*, 1955.
33. Gao Huang, Zhuang Liu, Laurens Van Der Maaten, and Kilian Q Weinberger. Densely connected convolutional networks. In *IEEE Conference on Computer Vision and Pattern Recognition*, 2017.
34. Subrata Chakraborty and Dhrubajyoti Chakravarty. A new discrete probability distribution with integer support on $(-\infty, \infty)$. *Communications in Statistics-Theory and Methods*, 45(2):492–505, 2016.
35. Tim Salimans, Andrej Karpathy, Xi Chen, and Diederik P Kingma. Pixelcnn++: Improving the PixelCNN with discretized logistic mixture likelihood and other modifications. *arXiv preprint arXiv:1701.05517*, 2017.
36. Emile van Krieken, Jakub M Tomczak, and Annette ten Teije. Storchastic: A framework for general stochastic automatic differentiation. *Advances in Neural Information Processing Systems*, 2021.
37. Niru Maheswaranathan, Luke Metz, George Tucker, Dami Choi, and Jascha Sohl-Dickstein. Guided evolutionary strategies: Augmenting random search with surrogate gradients. In *International Conference on Machine Learning*, pages 4264–4273. PMLR, 2019.

Chapter 4
Latent Variable Models

4.1 Introduction

In the previous chapters, we discussed two approaches to learning $p(\mathbf{x})$: autoregressive models (ARMs) in Chap. 2 and flow-based models (or flows for short) in Chap. 3. Both ARMs and flows model the likelihood function directly, that is, either by factorizing the distribution and parameterizing conditional distributions $p(x_d|\mathbf{x}_{<d})$ as in ARMs or by utilizing invertible transformations (neural networks) for the change of variables formula as in flows. Now, we will discuss a third approach that introduces **latent variables**.

Let us briefly discuss the following scenario. We have a collection of images with horses. We want to learn $p(\mathbf{x})$ for, e.g., generating new images. Before we do that, we can ask ourselves how we should generate a horse, or, in other words, if we were such a generative model, how we would do that. Maybe we would first sketch the general silhouette of a horse, its size and shape, then add hooves, fill in details of a head, color it, etc. In the end, we may consider the background. In general, we can say that there are some *factors* in data (e.g., a silhouette, a color, a background) that are crucial for generating an object (here, a horse). Once we decide about these factors, we can generate them by adding details. I do not want to delve into a philosophical/cognitive discourse, but I hope that we all agree that when we paint something, this is more-or-less our procedure of generating a painting.

We use mathematics now to express this *generative process*. Namely, we have our high-dimensional objects of interest, $\mathbf{x} \in \mathcal{X}^D$ (e.g., for images, $\mathcal{X} \in \{0, 1, \ldots, 255\}$), and **low-dimensional latent variables**, $\mathbf{z} \in \mathcal{Z}^M$ (e.g., $\mathcal{Z} = \mathbb{R}$), that we can call hidden factors in data. In mathematical words, we can refer to $\mathcal{Z}^M$ as a low-dimensional *manifold*. Then, the generative process could be expressed as follows:

1. $\mathbf{z} \sim p(\mathbf{z})$ (Fig. 4.1, in red)
2. $\mathbf{x} \sim p(\mathbf{x}|\mathbf{z})$ (Fig. 4.1, in blue)

J. M. Tomczak, *Deep Generative Modeling*,
https://doi.org/10.1007/978-3-030-93158-2_4

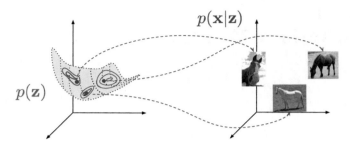

Fig. 4.1 A diagram presenting a latent variable model and a generative process. Notice the low-dimensional manifold (here 2D) embedded in the high-dimensional space (here 3D)

In plain words, we first sample $\mathbf{z}$ (e.g., we imagine the size, the shape, and the color of a horse) and then create an image with all necessary details, i.e., we sample $\mathbf{x}$ from the conditional distribution $p(\mathbf{x}|\mathbf{z})$. One can ask whether we need probabilities here but try to create *precisely the same* image at least two times. Due to various external factors, it is almost impossible to create two identical images. That is why probability theory is so beautiful and allows us to describe reality!

The idea behind **latent variable models** is that we introduce the latent variables $\mathbf{z}$ and the joint distribution is factorized as follows: $p(\mathbf{x}, \mathbf{z}) = p(\mathbf{x}|\mathbf{z})p(\mathbf{z})$. This naturally expressed the generative process described above. However, for training, we have access only to $\mathbf{x}$. Therefore, according to probabilistic inference, we should *sum out* (or *marginalize out*) the unknown, namely, $\mathbf{z}$. As a result, the (marginal) likelihood function is the following:

$$p(\mathbf{x}) = \int p(\mathbf{x}|\mathbf{z})p(\mathbf{z}) \, d\mathbf{z}. \qquad (4.1)$$

A natural question now is how to calculate this integral. In general, it is a difficult task. There are two possible directions. First, the integral **is** tractable. We will briefly discuss it before we jump into the second approach that utilizes a specific **approximate inference**, namely, **variational inference**.

4.2 Probabilistic Principal Component Analysis

Let us discuss the following situation:

- We consider continuous random variables only, i.e., $\mathbf{z} \in \mathbb{R}^M$ and $\mathbf{x} \in \mathbb{R}^D$.
- The distribution of $\mathbf{z}$ is the standard Gaussian, i.e., $p(\mathbf{z}) = \mathcal{N}(\mathbf{z}|0, \mathbf{I})$.
- The dependency between $\mathbf{z}$ and $\mathbf{x}$ is linear and we assume a Gaussian additive noise:

$$\mathbf{x} = \mathbf{W}\mathbf{z} + \mathbf{b} + \varepsilon, \qquad (4.2)$$

where $\varepsilon \sim \mathcal{N}(\varepsilon|0, \sigma^2 \mathbf{I})$. The property of the Gaussian distribution yields [1]

$$p(\mathbf{x}|\mathbf{z}) = \mathcal{N}\left(\mathbf{x}|\mathbf{W}\mathbf{z} + \mathbf{b}, \sigma^2 \mathbf{I}\right). \tag{4.3}$$

This model is known as the *probabilistic Principal Component Analysis* (pPCA) [2].

Next, we can take advantage of properties of a linear combination of two vectors of normally distributed random variables to calculate the integral explicitly [1]:

$$p(\mathbf{x}) = \int p(\mathbf{x}|\mathbf{z}) \, p(\mathbf{z}) \, d\mathbf{z} \tag{4.4}$$

$$= \int \mathcal{N}\left(\mathbf{x}|\mathbf{W}\mathbf{z} + \mathbf{b}, \sigma^2 \mathbf{I}\right) \mathcal{N}\left(\mathbf{z}|0, \mathbf{I}\right) \, d\mathbf{z} \tag{4.5}$$

$$= \mathcal{N}\left(\mathbf{x}|\mathbf{b}, \mathbf{W}\mathbf{W}^\top + \sigma^2 \mathbf{I}\right). \tag{4.6}$$

Now, we are able to calculate the logarithm of the (marginal) likelihood function $\ln p(\mathbf{x})$! We refer to [1, 2] for more details on learning the parameters in the pPCA model. Moreover, what is interesting about the pPCA is that, due to the properties of Gaussians, we can also calculate the *true* posterior over $\mathbf{z}$ analytically:

$$p(\mathbf{z}|\mathbf{x}) = \mathcal{N}\left(\mathbf{M}^{-1}\mathbf{W}^\top(\mathbf{x} - \mu), \sigma^{-2}\mathbf{M}\right), \tag{4.7}$$

where $\mathbf{M} = \mathbf{W}^\top \mathbf{W} + \sigma^2 \mathbf{I}$. Once we find $\mathbf{W}$ that maximize the log-likelihood function, and the dimensionality of the matrix $\mathbf{W}$ is computationally tractable, we can calculate $p(\mathbf{z}|\mathbf{x})$. This is a big thing! Why? Because for a given observation $\mathbf{x}$, we can calculate the distribution over the *latent factors*!

In my opinion, the probabilistic PCA is an extremely important latent variable model for two reasons. First, we can calculate everything *by hand* and, thus, it is a great exercise to develop an intuition about the latent variable models. Second, it is a linear model and, therefore, a curious reader should feel tingling in his or her head already and ask himself or herself the following questions: What would happen if we take non-linear dependencies? And what would happen if we use other distributions than Gaussians? In both cases, the answer is the same: We would not be able to calculate the integral exactly, and some sort of approximation would be necessary. Anyhow, pPCA is a model that everyone interested in latent variable models should study in depth to create an intuition about probabilistic modeling.

4.3   Variational Auto-Encoders: Variational Inference for Non-linear Latent Variable Models

4.3.1   The Model and the Objective

Let us take a look at the integral one more time and think of a general case where we cannot calculate it analytically. The simplest approach would be to use the Monte Carlo approximation:

$$p(\mathbf{x}) = \int p(\mathbf{x}|\mathbf{z})\, p(\mathbf{z})\, d\mathbf{z} \tag{4.8}$$

$$= \mathbb{E}_{\mathbf{z} \sim p(\mathbf{z})}\left[p(\mathbf{x}|\mathbf{z}) \right] \tag{4.9}$$

$$\approx \frac{1}{K} \sum_k p(\mathbf{x}|\mathbf{z}_k), \tag{4.10}$$

where, in the last line, we use samples from the prior over latents, $\mathbf{z}_k \sim p(\mathbf{z})$. Such an approach is relatively easy and since our computational power grows so fast, we can sample a lot of points in reasonably short time. However, as we know from statistics, if $\mathbf{z}$ is multidimensional, and M is relatively large, we get into a trap of the *curse of dimensionality*, and to cover the space properly, the number of samples grows exponentially with respect to M. If we take too few samples, then the approximation is simply very poor.

We can use more advanced Monte Carlo techniques [3]; however, they still suffer from issues associated with the curse of dimensionality. An alternative approach is the application of *variational inference* [4]. Let us consider a family of variational distributions parameterized by ϕ, $\{q_\phi(\mathbf{z})\}_\phi$. For instance, we can consider Gaussians with means and variances, $\phi = \{\mu, \sigma^2\}$. We know the form of these distributions, and we assume that they assign non-zero probability mass to all $\mathbf{z} \in \mathcal{Z}^M$. Then, the logarithm of the marginal distribution could be approximated as follows:

$$\ln p(\mathbf{x}) = \ln \int p(\mathbf{x}|\mathbf{z})p(\mathbf{z})\, d\mathbf{z} \tag{4.11}$$

$$= \ln \int \frac{q_\phi(\mathbf{z})}{q_\phi(\mathbf{z})} p(\mathbf{x}|\mathbf{z})p(\mathbf{z})\, d\mathbf{z} \tag{4.12}$$

$$= \ln \mathbb{E}_{\mathbf{z} \sim q_\phi(\mathbf{z})}\left[\frac{p(\mathbf{x}|\mathbf{z})p(\mathbf{z})}{q_\phi(\mathbf{z})} \right] \tag{4.13}$$

$$\geq \mathbb{E}_{\mathbf{z} \sim q_\phi(\mathbf{z})} \ln \left[\frac{p(\mathbf{x}|\mathbf{z})p(\mathbf{z})}{q_\phi(\mathbf{z})} \right] \tag{4.14}$$

$$= \mathbb{E}_{\mathbf{z} \sim q_\phi(\mathbf{z})} \left[\ln p(\mathbf{x}|\mathbf{z}) + \ln p(\mathbf{z}) - \ln q_\phi(\mathbf{z}) \right] \tag{4.15}$$

$$= \mathbb{E}_{\mathbf{z} \sim q_\phi(\mathbf{z})} \left[\ln p(\mathbf{x}|\mathbf{z}) \right] - \mathbb{E}_{\mathbf{z} \sim q_\phi(\mathbf{z})} \left[\ln q_\phi(\mathbf{z}) - \ln p(\mathbf{z}) \right]. \tag{4.16}$$

In the fourth line we used *Jensen's inequality*.

If we consider an *amortized variational posterior*, namely, $q_\phi(\mathbf{z}|\mathbf{x})$ instead of $q_\phi(\mathbf{z})$ for each $\mathbf{x}$, then we get

$$\ln p(\mathbf{x}) \geq \mathbb{E}_{\mathbf{z} \sim q_\phi(\mathbf{z}|\mathbf{x})} \left[\ln p(\mathbf{x}|\mathbf{z}) \right] - \mathbb{E}_{\mathbf{z} \sim q_\phi(\mathbf{z}|\mathbf{x})} \left[\ln q_\phi(\mathbf{z}|\mathbf{x}) - \ln p(\mathbf{z}) \right]. \quad (4.17)$$

Amortization could be extremely useful because we train a single model (e.g., a neural network with some weights), and it returns parameters of a distribution for given input. From now on, we will assume that we use amortized variational posteriors; however, please remember that we do not need to do that! Please take a look at [5] where a semi-amortized variational inference is considered.

As a result, we obtain an auto-encoder-like model, with a *stochastic encoder*, $q_\phi(\mathbf{z}|\mathbf{x})$, and a *stochastic decoder*, $p(\mathbf{x}|\mathbf{z})$. We use *stochastic* to highlight that the encoder and the decoder are probability distributions and to stress out a difference to a deterministic auto-encoder. This model, with the amortized variational posterior, is called a **Variational Auto-Encoder** [6, 7]. The lower bound of the log-likelihood function is called the Evidence Lower BOund (**ELBO**).

The first part of the ELBO, $\mathbb{E}_{\mathbf{z} \sim q_\phi(\mathbf{z}|\mathbf{x})} \left[\ln p(\mathbf{x}|\mathbf{z}) \right]$, is referred to as the (negative) *reconstruction error*, because $\mathbf{x}$ is encoded to $\mathbf{z}$ and then decoded back. The second part of the ELBO, $\mathbb{E}_{\mathbf{z} \sim q_\phi(\mathbf{z}|\mathbf{x})} \left[\ln q_\phi(\mathbf{z}|\mathbf{x}) - \ln p(\mathbf{z}) \right]$, could be seen as a *regularizer* and it coincides with the Kullback–Leibler (KL) divergence. Please keep in mind that for more complex models (e.g., hierarchical models), the regularizer(s) may not be interpreted as the KL term. Therefore, we prefer to use the term *the regularizer* because it is more general.

4.3.2 A Different Perspective on the ELBO

For completeness, we provide also a different derivation of the ELBO that will help us to understand why the lower bound might be tricky sometimes:

$$\ln p(\mathbf{x}) = \mathbb{E}_{\mathbf{z} \sim q_\phi(\mathbf{z}|\mathbf{x})} \left[\ln p(\mathbf{x}) \right] \quad (4.18)$$

$$= \mathbb{E}_{\mathbf{z} \sim q_\phi(\mathbf{z}|\mathbf{x})} \left[\ln \frac{p(\mathbf{z}|\mathbf{x}) p(\mathbf{x})}{p(\mathbf{z}|\mathbf{x})} \right] \quad (4.19)$$

$$= \mathbb{E}_{\mathbf{z} \sim q_\phi(\mathbf{z}|\mathbf{x})} \left[\ln \frac{p(\mathbf{x}|\mathbf{z}) p(\mathbf{z})}{p(\mathbf{z}|\mathbf{x})} \right] \quad (4.20)$$

$$= \mathbb{E}_{\mathbf{z} \sim q_\phi(\mathbf{z}|\mathbf{x})} \left[\ln \frac{p(\mathbf{x}|\mathbf{z}) p(\mathbf{z})}{p(\mathbf{z}|\mathbf{x})} \frac{q_\phi(\mathbf{z}|\mathbf{x})}{q_\phi(\mathbf{z}|\mathbf{x})} \right] \quad (4.21)$$

$$= \mathbb{E}_{\mathbf{z} \sim q_\phi(\mathbf{z}|\mathbf{x})} \left[\ln p(\mathbf{x}|\mathbf{z}) \frac{p(\mathbf{z})}{q_\phi(\mathbf{z}|\mathbf{x})} \frac{q_\phi(\mathbf{z}|\mathbf{x})}{p(\mathbf{z}|\mathbf{x})} \right] \quad (4.22)$$

Fig. 4.2 The ELBO is a lower bound on the log-likelihood. As a result, $\hat{\theta}$ maximizing the ELBO does not necessarily coincide with θ^* that maximizes $\ln p(\mathbf{x})$. The looser the ELBO is, the more this can bias maximum likelihood estimates of the model parameters

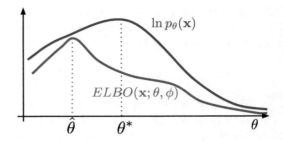

$$= \mathbb{E}_{\mathbf{z} \sim q_\phi(\mathbf{z}|\mathbf{x})} \left[\ln p(\mathbf{x}|\mathbf{z}) - \ln \frac{q_\phi(\mathbf{z}|\mathbf{x})}{p(\mathbf{z})} + \ln \frac{q_\phi(\mathbf{z}|\mathbf{x})}{p(\mathbf{z}|\mathbf{x})} \right] \tag{4.23}$$

$$= \mathbb{E}_{\mathbf{z} \sim q_\phi(\mathbf{z}|\mathbf{x})} \left[\ln p(\mathbf{x}|\mathbf{z}) \right] - KL \left[q_\phi(\mathbf{z}|\mathbf{x}) \| p(\mathbf{z}) \right] + KL \left[q_\phi(\mathbf{z}|\mathbf{x}) \| p(\mathbf{z}|\mathbf{x}) \right]. \tag{4.24}$$

Please note that in the derivation above we use the sum and the product rules together with multiplying by $1 = \frac{q_\phi(\mathbf{z}|\mathbf{x})}{q_\phi(\mathbf{z}|\mathbf{x})}$, nothing else, no dirty tricks here! Please try to replicate this by yourself, step by step. If you understand this derivation well, it would greatly help you to see where potential problems of the VAEs (and the latent variable models in general) lie.

Once you analyzed this derivation, let us take a closer look at it:

$$\ln p(\mathbf{x}) = \underbrace{\mathbb{E}_{\mathbf{z} \sim q_\phi(\mathbf{z}|\mathbf{x})} \left[\ln p(\mathbf{x}|\mathbf{z}) \right] - KL \left[q_\phi(\mathbf{z}|\mathbf{x}) \| p(\mathbf{z}) \right]}_{ELBO} + \underbrace{KL \left[q_\phi(\mathbf{z}|\mathbf{x}) \| p(\mathbf{z}|\mathbf{x}) \right]}_{\geq 0}.$$

$$(4.25)$$

The last component, $KL \left[q_\phi(\mathbf{z}|\mathbf{x}) \| p(\mathbf{z}|\mathbf{x}) \right]$, measures the difference between the variational posterior and the *real* posterior, but we do not know what the real posterior is! However, we can skip this part since the Kullback–Leibler divergence is always equal to or greater than 0 (from its definition) and, thus, we are left with the ELBO. We can think of $KL \left[q_\phi(\mathbf{z}|\mathbf{x}) \| p(\mathbf{z}|\mathbf{x}) \right]$ as a gap between the ELBO and the true log-likelihood.

Beautiful! But ok, why this is so important? Well, if we take $q_\phi(\mathbf{z}|\mathbf{x})$ that is a bad approximation of $p(\mathbf{z}|\mathbf{x})$, then the KL term will be larger, and even if the ELBO is optimized well, the gap between the ELBO and the true log-likelihood could be huge! In plain words, if we take too simplistic posterior, we can end up with a bad VAE anyway. What is "bad" in this context? Let us take a look at Fig. 4.2. If the ELBO is a loose lower bound of the log-likelihood, then the optimal solution of the ELBO could be completely different than the solution of the log-likelihood. We will comment on how to deal with that later on and, for now, it is enough to be aware of that issue.

4.3.3   Components of VAEs

Let us wrap up what we know right now. First of all, we consider a class of amortized variational posteriors $\{q_\phi(\mathbf{z}|\mathbf{x})\}_\phi$ that approximate the true posterior $p(\mathbf{z}|\mathbf{x})$. We can see them as **stochastic encoders**. Second, the conditional likelihood $p(\mathbf{x}|\mathbf{z})$ could be seen as a **stochastic decoder**. Third, the last component, $p(\mathbf{z})$, is the **marginal distribution**, also referred to as a **prior**. Lastly, the objective is the ELBO, a lower bound to the log-likelihood function:

$$\ln p(\mathbf{x}) \geq \mathbb{E}_{\mathbf{z}\sim q_\phi(\mathbf{z}|\mathbf{x})}\left[\ln p(\mathbf{x}|\mathbf{z})\right] - \mathbb{E}_{\mathbf{z}\sim q_\phi(\mathbf{z}|\mathbf{x})}\left[\ln q_\phi(\mathbf{z}|\mathbf{x}) - \ln p(\mathbf{z})\right]. \qquad (4.26)$$

There are two questions left to get the full picture of the VAEs:

1. How to parameterize the distributions?
2. How to calculate the expected values? After all, these integrals have not disappeared!

4.3.3.1   Parameterization of Distributions

As you can probably guess by now, we use neural networks to parameterize the encoders and the decoders. But before we use the neural networks, we should know *what* distributions we use! Fortunately, in the VAE framework we are almost free to choose any distributions! However, we must remember that they should make sense for a considered problem. So far, we have explained everything through images, so let us continue that. If $\mathbf{x} \in \{0, 1, \ldots, 255\}^D$, then we *cannot* use a Normal distribution, because its support is totally different than the support of discrete-valued images. A possible distribution we can use is the *categorical distribution*, that is:

$$p_\theta(\mathbf{x}|\mathbf{z}) = \text{Categorical}\left(\mathbf{x}|\theta(\mathbf{z})\right), \qquad (4.27)$$

where the probabilities are given by a neural network NN, namely, $\theta(\mathbf{z}) = \text{softmax}(\text{NN}(\mathbf{z}))$. The neural network NN could be an MLP, a convolutional neural network, RNNs, etc.

The choice of a distribution for the latent variables depends on how we want to express the latent factors in data. For convenience, typically $\mathbf{z}$ is taken as a vector of continuous random variables, $\mathbf{z} \in \mathbb{R}^M$. Then, we can use Gaussians for both the variational posterior and the prior:

$$q_\phi(\mathbf{z}|\mathbf{x}) = \mathcal{N}\left(\mathbf{z}|\mu_\phi(\mathbf{x}), \text{diag}\left[\sigma_\phi^2(\mathbf{x})\right]\right) \qquad (4.28)$$

$$p(\mathbf{z}) = \mathcal{N}\left(\mathbf{z}|0, \mathbf{I}\right), \qquad (4.29)$$

where $\mu_\phi(\mathbf{x})$ and $\sigma_\phi^2(\mathbf{x})$ are outputs of a neural network, similarly to the case of the decoder. In practice, we can have a shared neural network $NN(\mathbf{x})$ that outputs $2M$ values that are further split into M values for the mean μ and M values for the variance σ^2. For convenience, we consider a diagonal covariance matrix. We could use flexible posteriors (see Sect. 4.4.2). Moreover, here we take the standard Gaussian prior. We will comment on that later (see Sect. 4.4.1).

4.3.3.2   Reparameterization Trick

So far, we played around with the log-likelihood and we ended up with the ELBO. However, there is still a problem with calculating the expected value, because it contains an integral! Therefore, the question is how we can calculate it and why it is better than the MC-approximation of the log-likelihood without the variational posterior. In fact, we will use the MC-approximation, but now, instead of sampling from the prior $p(\mathbf{z})$, we will sample from the variational posterior $q_\phi(\mathbf{z}|\mathbf{x})$. Is it better? Yes, because the variational posterior assigns typically more probability mass to a smaller region than the prior. If you play around with your code of a VAE and examine the variance, you will probably notice that the variational posteriors are almost deterministic (whether it is good or bad is rather an open question). As a result, we should get a better approximation! However, there is still an issue with the variance of the approximation. Simply speaking, if we sample $\mathbf{z}$ from $q_\phi(\mathbf{z}|\mathbf{x})$, plug them into the ELBO, and calculate gradients with respect to the parameters of a neural network ϕ, the variance of the gradient may still be pretty large! A possible solution to that, first noticed by statisticians (e.g., see [8]), is the approach of **reparameterizing** the distribution. The idea is to realize that we can express a random variable as a composition of primitive transformations (e.g., arithmetic operations, logarithm, etc.) of an independent random variable with a simple distribution. For instance, if we consider a Gaussian random variable z with a mean μ and a variance σ^2, and an independent random variable $\epsilon \sim \mathcal{N}(\epsilon|0, 1)$, then the following holds (see Fig. 4.3):

$$z = \mu + \sigma \cdot \epsilon. \tag{4.30}$$

Now, if we start sampling ϵ from the standard Gaussian and apply the above transformation, then we get a sample from $\mathcal{N}(z|\mu, \sigma)$!

If you do not remember this fact from statistics, or you simply do not believe me, write a simple code for that and play around with it. In fact, this idea could be applied to many more distributions [9].

The **reparameterization trick** could be used in the encoder $q_\phi(\mathbf{z}|\mathbf{x})$. As observed by Kingma and Welling [6] and Rezende et al. [7], we can drastically reduce the variance of the gradient by using this reparameterization of the Gaussian distribution. Why? Because the randomness comes from the independent source $p(\epsilon)$, and we calculate gradient with respect to a deterministic function (i.e., a neural

Fig. 4.3 An example of reparameterizing a Gaussian distribution: We scale ϵ distributed according to the standard Gaussian by σ and shift it by μ

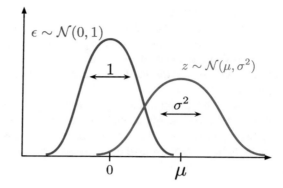

network), not random objects. Even better, since we learn the VAE using stochastic gradient descent, it is enough to sample $\mathbf{z}$ only once during training!

4.3.4   VAE in Action!

We went through a lot of theory and discussions, and you might think it is impossible to implement a VAE. However, it is actually simpler than it might look. Let us sum up what we know so far and focus on very specific distributions and neural networks.

First of all, we will use the following distributions:

- $q_\phi(\mathbf{z}|\mathbf{x}) = \mathcal{N}\left(\mathbf{z}|\mu_\phi(\mathbf{x}), \sigma_\phi^2(\mathbf{x})\right)$;
- $p(\mathbf{z}) = \mathcal{N}(\mathbf{z}|0, \mathbf{I})$;
- $p_\theta(\mathbf{x}|\mathbf{z}) = \text{Categorical}(\mathbf{x}|\theta(\mathbf{z}))$.

We assume that $x_d \in \mathcal{X} = \{0, 1, \ldots, L - 1\}$.

Next, we will use the following networks:

- The *encoder network*:

$$\mathbf{x} \in \mathcal{X}^D \rightarrow \text{Linear}(D, 256) \rightarrow \text{LeakyReLU} \rightarrow$$

$$\text{Linear}(256, 2 \cdot M) \rightarrow \text{split} \rightarrow \mu \in \mathbb{R}^M, \ \log \sigma^2 \in \mathbb{R}^M.$$

Notice that the last layer outputs $2M$ values because we must have M values for the mean and M values for the (log-)variance. Moreover, a variance must be positive; therefore, instead, we consider the logarithm of the variance because it can take real values then. As a result, we do not need to bother about variances being always positive. An alternative is to apply a non-linearity like softplus.
- The *decoder network*:

$$\mathbf{z} \in \mathbb{R}^M \rightarrow \text{Linear}(M, 256) \rightarrow \text{LeakyReLU} \rightarrow$$

$$\text{Linear}(256, D \cdot L) \rightarrow \text{reshape} \rightarrow \text{softmax} \rightarrow \theta \in [0, 1]^{D \times L}.$$

Since we use the categorical distribution for $\mathbf{x}$, the outputs of the decoder network are probabilities. Thus, the last layer must output $D \cdot L$ values, where D is the number of pixels and L is the number of possible values of a pixel. Then, we must reshape the output to a tensor of the following shape: (B, D, L), where B is the batch size. Afterward, we can apply the softmax activation function to obtain probabilities.

Finally, for a given dataset $\mathcal{D} = \{\mathbf{x}_n\}_{n=1}^N$, the training objective is the ELBO where we use a single sample from the variational posterior $\mathbf{z}_{\phi,n} = \mu_\phi(\mathbf{x}_n) + \sigma_\phi(\mathbf{x}_n) \odot \epsilon$. We must remember that in almost any available package we minimize by default, so we must take the negative sign, namely:

$$
-ELBO(\mathcal{D}; \theta, \phi) = \sum_{n=1}^N - \left\{ \ln \text{Categorical} \left(\mathbf{x}_n | \theta \left(\mathbf{z}_{\phi,n} \right) \right) + \right.
$$
$$
\left. \left[\ln \mathcal{N} \left(\mathbf{z}_{\phi,n} | \mu_\phi(\mathbf{x}_n), \sigma_\phi^2(\mathbf{x}_n) \right) + \ln \mathcal{N} \left(\mathbf{z}_{\phi,n} | 0, \mathbf{I} \right) \right] \right\}.
$$
(4.31)

So as you can see, the whole math boils down to a relatively simple learning procedure:

1. Take $\mathbf{x}_n$ and apply the encoder network to get $\mu_\phi(\mathbf{x}_n)$ and $\ln \sigma_\phi^2(\mathbf{x}_n)$.
2. Calculate $\mathbf{z}_{\phi,n}$ by applying the reparameterization trick, $\mathbf{z}_{\phi,n} = \mu_\phi(\mathbf{x}_n) + \sigma_\phi(\mathbf{x}_n) \odot \epsilon$, where $\epsilon \sim \mathcal{N}(0, \mathbf{I})$.
3. Apply the decoder network to $\mathbf{z}_{\phi,n}$ to get the probabilities $\theta(\mathbf{z}_{\phi,n})$.
4. Calculate the ELBO by plugging in $\mathbf{x}_n$, $\mathbf{z}_{\phi,n}$, $\mu_\phi(\mathbf{x}_n)$, and $\ln \sigma_\phi^2(\mathbf{x}_n)$.

4.3.5   Code

Now, all components are ready to be turned into a code! For the full implementation, please take a look at https://github.com/jmtomczak/intro_dgm. Here, we focus only on the code for the VAE model. We provide details in the comments. We divide the code into four classes: Encoder, Decoder, Prior, and VAE. It might look like overkill, but it may help you to think of the VAE as a composition of three parts and better comprehend the whole approach.

```
1  class Encoder(nn.Module):
2      def __init__(self, encoder_net):
3          super(Encoder, self).__init__()
4
5          # The init of the encoder network.
6          self.encoder = encoder_net
7
8          # The reparameterization trick for Gaussians.
9          @staticmethod
10         def reparameterization(mu, log_var):
```

```
11      # The formula is the following:
12      # z = mu + std * epsilon
13      # epsilon ~ Normal(0,1)
14
15      # First, we need to get std from log-variance.
16      std = torch.exp(0.5*log_var)
17
18      # Second, we sample epsilon from Normal(0,1).
19      eps = torch.randn_like(std)
20
21      # The final output
22      return mu + std * eps
23
24  # This function implements the output of the encoder network
    (i.e., parameters of a Gaussian).
25  def encode(self, x):
26      # First, we calculate the output of the encoder network
    of size 2M.
27      h_e = self.encoder(x)
28      # Second, we must divide the output to the mean and the
    log-variance.
29      mu_e, log_var_e = torch.chunk(h_e, 2, dim=1)
30      return mu_e, log_var_e
31
32  # Sampling procedure.
33  def sample(self, x=None, mu_e=None, log_var_e=None):
34      #If we don't provide a mean and a log-variance, we must
    first calculate it:
35      if (mu_e is None) and (log_var_e is None):
36          mu_e, log_var_e = self.encode(x)
37      # Or the final sample
38      else:
39      # Otherwise, we can simply apply the reparameterization
    trick!
40          if (mu_e is None) or (log_var_e is None):
41              raise ValueError('mu and log-var can't be None!')
42      z = self.reparameterization(mu_e, log_var_e)
43      return z
44
45  # This function calculates the log-probability that is later
    used for calculating the ELBO.
46  def log_prob(self, x=None, mu_e=None, log_var_e=None, z=None)
    :
47      # If we provide x alone, then we can calculate a
    corresponding sample:
48      if x is not None:
49          mu_e, log_var_e = self.encode(x)
50          z = self.sample(mu_e=mu_e, log_var_e=log_var_e)
51      else:
52      # Otherwise, we should provide mu, log-var and z!
53          if (mu_e is None) or (log_var_e is None) or (z is
    None):
54              raise ValueError('mu, log-var, z can't be None')
55
```

```
56        return log_normal_diag(z, mu_e, log_var_e)
57
58    # PyTorch forward pass: it is either log-probability (by
      default) or sampling.
59    def forward(self, x, type='log_prob'):
60        assert type in ['encode', 'log_prob'], 'Type could be
      either encode or log_prob'
61        if type == 'log_prob':
62            return self.log_prob(x)
63        else:
64            return self.sample(x)
```

Listing 4.1 An encoder class

```
1  class Decoder(nn.Module):
2      def __init__(self, decoder_net, distribution='categorical',
        num_vals=None):
3          super(Decoder, self).__init__()
4
5          # The decoder network.
6          self.decoder = decoder_net
7          # The distribution used for the decoder (it is
        categorical by default, as discussed above).
8          self.distribution = distribution
9          # The number of possible values. This is important for
        the categorical distribution.
10         self.num_vals=num_vals
11
12     # This function calculates parameters of the likelihood
       function p(x|z)
13     def decode(self, z):
14         # First, we apply the decoder network.
15         h_d = self.decoder(z)
16
17         # In this example, we use only the categorical
       distribution...
18         if self.distribution == 'categorical':
19             # We save the shapes: batch size
20             b = h_d.shape[0]
21             # and the dimensionality of x.
22             d = h_d.shape[1]//self.num_vals
23             # Then we reshape to (Batch size, Dimensionality,
       Number of Values).
24             h_d = h_d.view(b, d, self.num_vals)
25             # To get probabilities, we apply softmax.
26             mu_d = torch.softmax(h_d, 2)
27             return [mu_d]
28         # ... however, we also present the Bernoulli distribution
       . We are nice, aren't we?
29         elif self.distribution == 'bernoulli':
30             # In the Bernoulli case, we have x_d \in {0,1}.
       Therefore, it is enough to output a single probability,
31             # because p(x_d=1|z) = \theta and p(x_d=0|z) = 1 - \
       theta
```

```
32            mu_d = torch.sigmoid(h_d)
33            return [mu_d]
34
35        else:
36            raise ValueError('Only: 'categorical', 'bernoulli'')
37
38    # This function implements sampling from the decoder.
39    def sample(self, z):
40        outs = self.decode(z)
41
42        if self.distribution == 'categorical':
43            # We take the output of the decoder
44            mu_d = outs[0]
45            # and save shapes (we will need that for reshaping).
46            b = mu_d.shape[0]
47            m = mu_d.shape[1]
48            # Here we use reshaping
49            mu_d = mu_d.view(mu_d.shape[0], -1, self.num_vals)
50            p = mu_d.view(-1, self.num_vals)
51            # Eventually, we sample from the categorical (the
built-in PyTorch function).
52            x_new = torch.multinomial(p, num_samples=1).view(b,m)
53
54        elif self.distribution == 'bernoulli':
55            # In the case of Bernoulli, we don't need any
reshaping
56            mu_d = outs[0]
57            # and we can use the built-in PyTorch sampler!
58            x_new = torch.bernoulli(mu_d)
59
60        else:
61            raise ValueError('Only: 'categorical', 'bernoulli'')
62
63        return x_new
64
65    # This function calculates the conditional log-likelihood
function.
66    def log_prob(self, x, z):
67        outs = self.decode(z)
68
69        if self.distribution == 'categorical':
70            mu_d = outs[0]
71            log_p = log_categorical(x, mu_d, num_classes=self.
num_vals, reduction='sum', dim=-1).sum(-1)
72
73        elif self.distribution == 'bernoulli':
74            mu_d = outs[0]
75            log_p = log_bernoulli(x, mu_d, reduction='sum', dim
=-1)
76
77        else:
78            raise ValueError('Only: 'categorical', 'bernoulli'')
79
80        return log_p
```

```
81
82      # The forward pass is either a log-prob or a sample.
83      def forward(self, z, x=None, type='log_prob'):
84          assert type in ['decoder', 'log_prob'], 'Type could be
        either decode or log_prob'
85          if type == 'log_prob':
86              return self.log_prob(x, z)
87          else:
88              return self.sample(x)
```

Listing 4.2 A decoder class

```
1 # The current implementation of the prior is very simple, namely,
      it is a standard Gaussian.
2 # We could have used a built-in PyTorch distribution. However, we
      didn't do that for two reasons:
3 # (i) It is important to think of the prior as a crucial
      component in VAEs.
4 # (ii) We can implement a learnable prior (e.g., a flow-based
      prior, VampPrior, a mixture of distributions).
5 class Prior(nn.Module):
6     def __init__(self, L):
7         super(Prior, self).__init__()
8         self.L = L
9
10    def sample(self, batch_size):
11        z = torch.randn((batch_size, self.L))
12        return z
13
14    def log_prob(self, z):
15        return log_standard_normal(z)
```

Listing 4.3 A prior class

```
1 class VAE(nn.Module):
2     def __init__(self, encoder_net, decoder_net, num_vals=256, L
      =16, likelihood_type='categorical'):
3         super(VAE, self).__init__()
4
5         print('VAE by JT.')
6
7         self.encoder = Encoder(encoder_net=encoder_net)
8         self.decoder = Decoder(distribution=likelihood_type,
      decoder_net=decoder_net, num_vals=num_vals)
9         self.prior = Prior(L=L)
10
11        self.num_vals = num_vals
12
13        self.likelihood_type = likelihood_type
14
15    def forward(self, x, reduction='avg'):
16        # encoder
17        mu_e, log_var_e = self.encoder.encode(x)
18        z = self.encoder.sample(mu_e=mu_e, log_var_e=log_var_e)
```

```
19
20          # ELBO
21          RE = self.decoder.log_prob(x, z)
22          KL = (self.prior.log_prob(z) - self.encoder.log_prob(mu_e
            =mu_e, log_var_e=log_var_e, z=z)).sum(-1)
23
24          if reduction == 'sum':
25              return -(RE + KL).sum()
26          else:
27              return -(RE + KL).mean()
28
29      def sample(self, batch_size=64):
30          z = self.prior.sample(batch_size=batch_size)
31          return self.decoder.sample(z)
```

Listing 4.4 A VAE class

```
1  # Examples of neural networks used for parameterizing the encoder
       and the decoder.
2
3  # Remember that the encoder outputs 2 times more values because
       we need L means and L log-variances for a Gaussian.
4  encoder = nn.Sequential(nn.Linear(D, M), nn.LeakyReLU(),
5                          nn.Linear(M, M), nn.LeakyReLU(),
6                          nn.Linear(M, 2 * L))
7
8  # Here we must remember that if we use the categorical
       distribution, we must output num_vals per each pixel.
9  decoder = nn.Sequential(nn.Linear(L, M), nn.LeakyReLU(),
10                          nn.Linear(M, M), nn.LeakyReLU(),
11                          nn.Linear(M, num_vals * D))
```

Listing 4.5 Examples of networks

Perfect! Now we are ready to run the full code and after training our VAE, we should obtain results similar to those like in Fig. 4.4.

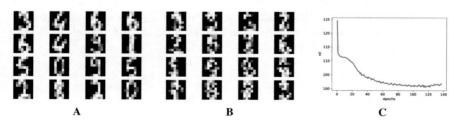

| A | B | C |

Fig. 4.4 An example of outcomes after the training: (**a**) Randomly selected real images. (**b**) Unconditional generations from the VAE. (**c**) The validation curve during training

4.3.6   Typical Issues with VAEs

VAEs constitute a very powerful class of models, mainly due to their flexibility. Unlike flow-based models, they do not require the invertibility of neural networks and, thus, we can use any arbitrary architecture for encoders and decoders. In contrast to ARMs, they learn a low-dimensional data representation and we can control the bottleneck (i.e., the dimensionality of the latent space). However, they also suffer from several issues. Except the ones mentioned before (i.e., a necessity of an efficient integral estimation, a gap between the ELBO and the log-likelihood function for too simplistic variational posteriors), the potential problems are the following:

- Let us take a look at the ELBO and the regularization term. For a non-trainable prior like the standard Gaussian, the regularization term will be minimized if $\forall_{\mathbf{x}} q_\phi(\mathbf{z}|\mathbf{x}) = p(\mathbf{z})$. This may happen if the decoder is so powerful that it treats $\mathbf{z}$ as a noise, e.g., a decoder is expressed by an ARM [10]. This issue is known as the *posterior collapse* [11].
- Another issue is associated with a mismatch between the aggregated posterior, $q_\phi(\mathbf{z}) = \frac{1}{N} \sum_n q_\phi(\mathbf{z}|\mathbf{x}_n)$, and the prior $p(\mathbf{z})$. Imagine that we have the standard Gaussian prior and the aggregated posterior (i.e., an average of variational posteriors over all training data). As a result, there are regions where there prior assigns high probability but the aggregated posterior assigns low probability, or other way around. Then, sampling from these holes provides unrealistic latent values and the decoder produces images of very low quality. This problem is referred to as the *hole problem* [12].
- The last problem we want to discuss is more general and, in fact, it affects all deep generative models. As it was noticed in [13], the deep generative models (including VAEs) fail to properly detect out-of-distribution examples. Out-of-distribution datapoints are examples that follow a totally different distribution than the one a model was trained on. For instance, let us assume that our model is trained on MNIST, then Fashion MNIST examples are out-of-distribution. Thus, an intuition tells that a properly trained deep generative model should assign high probability to in-distribution examples and low probability to out-of-distribution points. Unfortunately, as shown in [13], this is not the case. The *out-of-distribution problem* remains one of the main unsolved problems in deep generative modeling [14].

4.3.7   There Is More!

There are a plethora of papers that extend VAEs and apply them to many problems. Below, we will list out selected papers and only touch upon the vast literature on the topic!

Estimation of the Log-Likelihood Using Importance Weighting As we indicated multiple times, the ELBO is the lower bound to the log-likelihood and it rather should not be used as a good estimate of the log-likelihood. In [7, 15], an *importance weighting* procedure is advocated to better approximate the log-likelihood, namely:

$$\ln p(\mathbf{x}) \approx \ln \frac{1}{K} \sum_{k=1}^{K} \frac{p(\mathbf{x}|\mathbf{z}_k)}{q_\phi(\mathbf{z}_k|\mathbf{x})}, \tag{4.32}$$

where $\mathbf{z}_k \sim q_\phi(\mathbf{z}_k|\mathbf{x})$. Notice that the logarithm is **outside** the expected value. As shown in [15], using importance weighting with sufficiently large K gives a good estimate of the log-likelihood. In practice, K is taken to be 512 or more if the computational budget allows.

Enhancing VAEs: Better Encoders After introducing the idea of VAEs, many papers focused on proposing a flexible family of variational posteriors. The most prominent direction is based on utilizing conditional flow-based models [16–21]. We discuss this topic more in Sect. 4.4.2.

Enhancing VAEs: Better Decoders VAEs allow using any neural network to parameterize the decoder. Therefore, we can use fully connected networks, fully convolutional networks, ResNets, or ARMs. For instance, in [22], a PixelCNN-based decoder was used utilized in a VAE.

Enhancing VAEs: Better Priors As mentioned before, this could be a serious issue if there is a big mismatch between the aggregated posterior and the prior. There are many papers that try to alleviate this issue by using a multimodal prior mimicking the aggregated posterior (known as the VampPrior) [23], or a flow-based prior (e.g., [24, 25]), an ARM-based prior [26], or using an idea of resampling [27]. We present various priors in Sect. 4.4.1.

Extending VAEs Here, we present the unsupervised version of VAEs. However, there is no restriction to that and we can introduce labels or other variables. In [28] a semi-supervised VAE was proposed. This idea was further extended to the concept of fair representations [29, 30]. In [30], the authors proposed a specific latent representation that allows domain generalization in VAEs. In [31] variational inference and the reparameterization trick were used for Bayesian Neural Nets. [31] is not necessarily introducing a VAE, but a VAE-like way of dealing with Bayesian neural nets.

VAEs for Non-image Data So far, we explain everything on images. However, there is no restriction on that! In [11] a VAE was proposed to deal with sequential data (e.g., text). The encoder and the decoder were parameterized by LSTMs. An interesting application of the VAE framework was also presented in [32] where VAEs were used for the molecular graph generation. In [26] the authors proposed a VAE-like model for video compression.

Different Latent Spaces Typically, the Euclidean latent space is considered. However, the VAE framework allows us to think of other spaces. For instance, in [33, 34] a hyperspherical latent space was used, and in [35] the hyperbolic latent space was utilized. More details about hyperspherical VAEs could be found in Sect. 4.4.2.3.

Discrete Latent Spaces We discuss the VAE framework with continuous latent variables. However, an interesting question is how to deal with discrete latent variables. The problem here is that we cannot use the reparameterization trick anymore. There are two potential solutions to that. First, a relaxation to the discrete variables could be used like the *Gumbel-Softmax trick* [36, 37]. Second, a method for a gradient estimation could be used [38].

The Posterior Collapse There were many ideas proposed to deal with the posterior collapse. For instance, [39] proposes to update variational posteriors more often than the decoder. In [40] a new architecture of the decoder is proposed by introducing *skip connections* to allow a better flow of information (thus, the gradients) in the decoder.

Various Perspectives on the Objective The core of the VAE is the ELBO. However, we can consider different objectives. For instance, [41] proposes an upper bound to the log-likelihood that is based on the chi-square divergence (CUBO). In [10] an information-theoretic perspective on the ELBO is presented. [42] introduced the β-VAE where the regularization term is weighted by a fudge factor β. The objective does not correspond to the lower bound of the log-likelihood though.

Deterministic Regularized Auto-Encoders We can take a look at the VAE and the objective, as mentioned before, and think of it as a regularized version of an auto-encoder with a stochastic encoder and a stochastic decoder. [43] "peeled off" VAEs from all stochasticity and indicated similarities between deterministic regularized auto-encoders and VAEs and highlighted potential issues with VAEs. Moreover, they brilliantly pointed out that even with a deterministic encoder, due to the stochasticity of the empirical distribution, we can fit a model to the aggregated posterior. As a result, the deterministic (regularized) auto-encoder could be turned into a generative model by sampling from a model of the aggregated posterior, $p_\lambda(\mathbf{z})$, and then, deterministically, mapping $\mathbf{z}$ to the space of observable $\mathbf{x}$. In my opinion, this direction should be further explored and an important question is whether we indeed need any stochasticity at all.

Hierarchical VAEs Very recently, there are many VAEs with a deep, hierarchical structure of latent variables that achieved remarkable results! The most important ones are definitely BIVA [44], NVAE [45], and very deep VAEs [46]. Another interesting perspective on a deep, hierarchical VAE was presented in [25] where, additionally, a series of deterministic functions was used. We delve into that topic in Sect. 4.5.

Adversarial Auto-Encoders Another interesting perspective on VAEs is presented in [47]. Since learning the aggregated posterior as the prior is an important component mentioned in some papers (e.g., [23, 48]), a different approach would be to train the prior with an adversarial loss. Further, [47] present various ideas how auto-encoders could benefit from adversarial learning.

4.4 Improving Variational Auto-Encoders

4.4.1 Priors

Insights from Rewriting the ELBO
One of the crucial components of VAEs is the marginal distribution over $\mathbf{z}$'s. Now, we will take a closer look at this distribution, also called the *prior*. Before we start thinking about improving it, we inspect the ELBO one more time. We can write ELBO as follows:

$$\mathbb{E}_{\mathbf{x} \sim p_{data}(\mathbf{x})}[\ln p(\mathbf{x})] \geq \mathbb{E}_{\mathbf{x} \sim p_{data}(\mathbf{x})}\left[\mathbb{E}_{q_\phi(\mathbf{z}|\mathbf{x})}\left[\ln p_\theta(\mathbf{x}|\mathbf{z}) + \ln p_\lambda(\mathbf{z}) - \ln q_\phi(\mathbf{z}|\mathbf{x})\right]\right], \tag{4.33}$$

where we explicitly highlight the summation over training data, namely, the expected value with respect to $\mathbf{x}$'s from the empirical distribution $p_{data}(\mathbf{x}) = \frac{1}{N}\sum_{n=1}^{N}\delta(\mathbf{x} - \mathbf{x}_n)$, and $\delta(\cdot)$ is the Dirac delta.

The ELBO consists of two parts, namely, the reconstruction error:

$$RE \triangleq \mathbb{E}_{\mathbf{x} \sim p_{data}(\mathbf{x})}\left[\mathbb{E}_{q_\phi(\mathbf{z}|\mathbf{x})}\left[\ln p_\theta(\mathbf{x}|\mathbf{z})\right]\right], \tag{4.34}$$

and the regularization term between the encoder and the prior:

$$\Omega \triangleq \mathbb{E}_{\mathbf{x} \sim p_{data}(\mathbf{x})}\left[\mathbb{E}_{q_\phi(\mathbf{z}|\mathbf{x})}\left[\ln p_\lambda(\mathbf{z}) - \ln q_\phi(\mathbf{z}|\mathbf{x})\right]\right]. \tag{4.35}$$

Further, let us play a little bit with the regularization term Ω:

$$\Omega = \mathbb{E}_{\mathbf{x} \sim p_{data}(\mathbf{x})}\left[\mathbb{E}_{q_\phi(\mathbf{z}|\mathbf{x})}\left[\ln p_\lambda(\mathbf{z}) - \ln q_\phi(\mathbf{z}|\mathbf{x})\right]\right] \tag{4.36}$$

$$= \int p_{data}(\mathbf{x}) \int q_\phi(\mathbf{z}|\mathbf{x})\left[\ln p_\lambda(\mathbf{z}) - \ln q_\phi(\mathbf{z}|\mathbf{x})\right]d\mathbf{z}\,d\mathbf{x} \tag{4.37}$$

$$= \iint p_{data}(\mathbf{x})q_\phi(\mathbf{z}|\mathbf{x})\left[\ln p_\lambda(\mathbf{z}) - \ln q_\phi(\mathbf{z}|\mathbf{x})\right]d\mathbf{z}\,d\mathbf{x} \tag{4.38}$$

$$= \iint \frac{1}{N}\sum_{n}\delta(\mathbf{x} - \mathbf{x}_n)q_\phi(\mathbf{z}|\mathbf{x})\left[\ln p_\lambda(\mathbf{z}) - \ln q_\phi(\mathbf{z}|\mathbf{x})\right]d\mathbf{z}\,d\mathbf{x} \tag{4.39}$$

Fig. 4.5 An example of the aggregated posterior. Individual points are encoded as Gaussians in the 2D latent space (magenta), and the mixture of variational posteriors (the aggregated posterior) is presented by contours

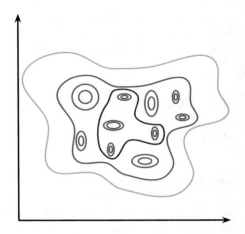

$$= \int \frac{1}{N} \sum_{n=1}^{N} q_\phi(\mathbf{z}|\mathbf{x}_n) \left[\ln p_\lambda(\mathbf{z}) - \ln q_\phi(\mathbf{z}|\mathbf{x}_n) \right] d\mathbf{z} \tag{4.40}$$

$$= \int \frac{1}{N} \sum_{n=1}^{N} q_\phi(\mathbf{z}|\mathbf{x}_n) \ln p_\lambda(\mathbf{z}) d\mathbf{z} - \int \frac{1}{N} \sum_{n=1}^{N} q_\phi(\mathbf{z}|\mathbf{x}_n) \ln q_\phi(\mathbf{z}|\mathbf{x}_n) d\mathbf{z} \tag{4.41}$$

$$= \int q_\phi(\mathbf{z}) \ln p_\lambda(\mathbf{z}) d\mathbf{z} - \int \sum_{n=1}^{N} \frac{1}{N} q_\phi(\mathbf{z}|\mathbf{x}_n) \ln q_\phi(\mathbf{z}|\mathbf{x}_n) d\mathbf{z} \tag{4.42}$$

$$= -\mathbb{CE} \left[q_\phi(\mathbf{z}) || p_\lambda(\mathbf{z}) \right] + \mathbb{H} \left[q_\phi(\mathbf{z}|\mathbf{x}) \right], \tag{4.43}$$

where we use the property of the Dirac delta: $\int \delta(a - a') f(a) da = f(a')$, and we use the notion of the **aggregated posterior** [47, 48] defined as follows:

$$q(\mathbf{z}) = \frac{1}{N} \sum_{n=1}^{N} q_\phi(\mathbf{z}|\mathbf{x}_n). \tag{4.44}$$

An example of the aggregated posterior is schematically depicted in Fig. 4.5.

Eventually, we obtain two terms:

(i) The first one, $\mathbb{CE} \left[q_\phi(\mathbf{z}) || p_\lambda(\mathbf{z}) \right]$, is the cross-entropy between the aggregated posterior and the prior.
(ii) The second term, $\mathbb{H} \left[q_\phi(\mathbf{z}|\mathbf{x}) \right]$, is the conditional entropy of $q_\phi(\mathbf{z}|\mathbf{x})$ with the empirical distribution $p_{data}(\mathbf{x})$.

I highly recommend doing this derivation step-by-step, as it helps a lot in understanding what is going on here. Interestingly, there is another possibility to rewrite Ω using three terms, with the total correlation [49]. We will not use it here, so it is left as a "homework."

Anyway, one may ask why is it useful to rewrite the ELBO? The answer is rather straightforward: We can analyze it from a different perspective! In this section, we will focus on the **prior**, an important component in the generative part that is very often neglected. Many Bayesianists are stating that a prior should not be learned. But VAEs are not Bayesian models, please remember that! Besides, who says we cannot learn the prior? As we will see shortly, a non-learnable prior could be pretty annoying, especially for the generation process.

What Does ELBO Tell Us About the Prior?

Alright, we see that Ω consists of the cross-entropy and the entropy. Let us start with the entropy since it is easier to be analyzed. While optimizing, we want to maximize the ELBO and, hence, we maximize the entropy:

$$\mathbb{H}\left[q_\phi(\mathbf{z}|\mathbf{x})\right] = -\int \sum_{n=1}^{N} \frac{1}{N} q_\phi(\mathbf{z}|\mathbf{x}_n) \ln q_\phi(\mathbf{z}|\mathbf{x}_n) \mathrm{d}\mathbf{z}. \tag{4.45}$$

Before we make any conclusions, we should remember that we consider Gaussian encoders, $q_\phi(\mathbf{z}|\mathbf{x}) = \mathcal{N}\left(\mathbf{z}|\mu(\mathbf{x}), \sigma^2(\mathbf{x})\right)$. The entropy of a Gaussian distribution with a diagonal covariance matrix is equal to $\frac{1}{2}\sum_i \ln\left(2e\pi\sigma_i^2\right)$. Then, the question is when this quantity is maximized? The answer is: $\sigma_i^2 \to +\infty$. In other words, the entropy terms tries to stretch the encoders as much as possible by enlarging their variances! Of course, this does not happen in practice because we use the encoder together with the decoder in the RE term and the decoder tries to make the encoder as peaky as possible (i.e., ideally one $\mathbf{x}$ for one $\mathbf{z}$, like in the non-stochastic auto-encoder).

The second term in Ω is the cross-entropy:

$$\mathbb{CE}\left[q_\phi(\mathbf{z})||p_\lambda(\mathbf{z})\right] = -\int q_\phi(\mathbf{z}) \ln p_\lambda(\mathbf{z}) \mathrm{d}\mathbf{z}. \tag{4.46}$$

The cross-entropy term influences the VAE in a different manner. First, we can ask the question how to interpret the cross-entropy between $q_\phi(\mathbf{z})$ and $p_\lambda(\mathbf{z})$. In general, the cross-entropy tells us the average number of bits (or rather nats because we use the natural logarithm) needed to identify an event drawn from $q_\phi(\mathbf{z}$ if a coding scheme used for it is $p_\lambda(\mathbf{z})$. Notice that in Ω we have the negative cross-entropy. Since we maximize the ELBO, it means that we aim for minimizing $\mathbb{CE}\left[q_\phi(\mathbf{z})||p_\lambda(\mathbf{z})\right]$. This makes sense because we would like $q_\phi(\mathbf{z})$ to match $p_\lambda(\mathbf{z})$. And we have accidentally touched upon the most important issue here: What do we really want here? The cross-entropy forces the aggregated posterior to **match** the prior! That is the reason why we have this term here. If you think about it, it is a beautiful result that gives another connection between VAEs and the information theory.

Fig. 4.6 An example of the effect of the cross-entropy optimization with a non-learnable prior. The aggregated posterior (purple contours) tries to match the non-learnable prior (in blue). The purple arrows indicate the change of the aggregated posterior. An example of a hole is presented as a dark gray ellipse

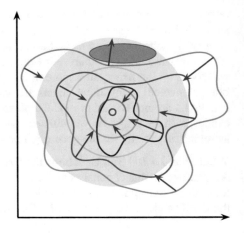

Alright, so we see what the cross-entropy does, but there are two possibilities here. First, the prior is fixed (**non-learnable**), e.g., the standard Gaussian prior. Then, optimizing the cross-entropy *pushes* the aggregated posterior to match the prior. It is schematically depicted in Fig. 4.6. The prior acts like an anchor and the *amoeba* of the aggregated posterior moves so that to fit the prior. In practice, this optimization process is troublesome because the decoder forces the encoder to be peaked and, in the end, it is almost impossible to match a fixed-shaped prior. As a result, we obtain **holes**, namely, regions in the latent space where the aggregated posterior assigns low probability while the prior assigns (relatively) high probability (see a dark gray ellipse in Fig. 4.6). This issue is especially apparent in generations because sampling from the prior, from the hole, may result in a sample that is of an extremely low quality. You can read more about it in [12].

On the other hand, if we consider a learnable **prior**, the situation looks different. The optimization allows to change the aggregated posterior **and** the prior. As the consequence, both distributions try to match each other (see Fig. 4.7). The problem of holes is then less apparent, especially if the prior is flexible enough. However, we can face other optimization issues when the prior and the aggregated posteriors chase each other. In practice, the learnable prior seems to be a better option, but it is still an open question whether training all components at once is the best approach. Moreover, the learnable prior does not impose any specific constraint on the latent representation, e.g., sparsity. This could be another problem that would result in undesirable problems (e.g., non-smooth encoders).

Eventually, we can ask the fundamental question: What is the *best* prior then?! The answer is already known and is hidden in the cross-entropy term: It is the aggregated posterior. If we take $p_\lambda(\mathbf{z}) = \sum_{n=1}^{N} \frac{1}{N} q_\phi(\mathbf{z}|\mathbf{x}_n)$, then, theoretically, the cross-entropy equals the entropy of $q_\phi(\mathbf{z})$ and the regularization term Ω is smallest. However, in practice, this is infeasible because:

- We cannot sum over tens of thousands of points and backpropagate through them.

Fig. 4.7 An example of the effect of the cross-entropy optimization with a learnable prior. The aggregated posterior (purple contours) tries to match the learnable prior (blue contours). Notice that the aggregated posterior is modified to fit the prior (purple arrows), but also the prior is updated to cover the aggregated posterior (orange arrows)

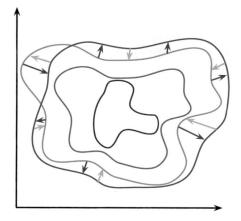

- This result is fine from the theoretical point of view; however, the optimization process is stochastic and could cause additional errors.
- As mentioned earlier, choosing the aggregated posterior as the prior does not constrain the latent representation in any obvious manner and, thus, the encoder could behave unpredictably.
- The aggregated posterior may work well if the get $N \to +\infty$ points, because then we can get any distribution; however, this is not the case in practice and it contradicts also the first bullet.

As a result, we can keep this theoretical solution in mind and formulate **approximations** to it that are computationally tractable. In the next sections, we will discuss a few of them.

4.4.1.1 Standard Gaussian

The vanilla implementation of the VAE assumes a standard Gaussian marginal (prior) over $\mathbf{z}$, $p_\lambda(\mathbf{z}) = \mathcal{N}(\mathbf{z}|0, \mathrm{I})$. This prior is simple, non-trainable (i.e., no extra parameters to learn), and easy to implement. In other words, it is amazing! However, as discussed above, the standard normal distribution could lead to very poor hidden representations with holes resulting from the mismatch between the aggregated posterior and the prior.

To strengthen our discussion, we trained a small VAE with the standard Gaussian prior and a two-dimensional latent space. In Fig. 4.8, we present samples from the encoder for the test data (black dots) and the contour plot for the standard prior. We can spot holes where the aggregated posterior does not assign any points (i.e., the mismatch between the prior and the aggregated posterior).

Fig. 4.8 An example of the
standard Gaussian prior
(contours) and the samples
from the aggregated posterior
(black dots)

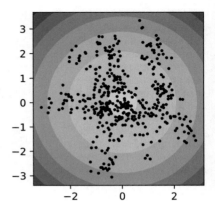

The code for the standard Gaussian prior is presented below:

```
class StandardPrior(nn.Module):
    def __init__(self, L=2):
        super(StandardPrior, self).__init__()

        self.L = L

        # params weights
        self.means = torch.zeros(1, L)
        self.logvars = torch.zeros(1, L)

    def get_params(self):
        return self.means, self.logvars

    def sample(self, batch_size):
        return torch.randn(batch_size, self.L)

    def log_prob(self, z):
        return log_standard_normal(z)
```

Listing 4.6 A standard Gaussian prior class

4.4.1.2   Mixture of Gaussians

If we take a closer look at the aggregated posterior, we immediately notice that it is
a mixture model, and a mixture of Gaussians, to be more precise. Therefore, we can
use the Mixture of Gaussians (MoG) prior with K components:

$$p_\lambda(\mathbf{z}) = \sum_{k=1}^{K} w_k \mathcal{N}(\mathbf{z}|\mu_k, \sigma_k^2), \tag{4.47}$$

where $\lambda = \{\{w_k\}, \{\mu_k\}, \{\sigma_k^2\}\}$ are trainable parameters.

Similarly to the standard Gaussian prior, we trained a small VAE with the mixture of Gaussians prior (with $K = 16$) and a two-dimensional latent space. In Fig. 4.9, we present samples from the encoder for the test data (black dots) and the contour plot for the MoG prior. Comparing to the standard Gaussian prior, the MoG prior fits better the aggregated posterior, allowing to *patch* holes.

An example of the code is presented below:

```
class MoGPrior(nn.Module):
    def __init__(self, L, num_components):
        super(MoGPrior, self).__init__()

        self.L = L
        self.num_components = num_components

        # params
        self.means = nn.Parameter(torch.randn(num_components,
    self.L)*multiplier)
        self.logvars = nn.Parameter(torch.randn(num_components,
    self.L))

        # mixing weights
        self.w = nn.Parameter(torch.zeros(num_components, 1, 1))

    def get_params(self):
        return self.means, self.logvars

    def sample(self, batch_size):
        # mu, lof_var
        means, logvars = self.get_params()

        # mixing probabilities
        w = F.softmax(self.w, dim=0)
        w = w.squeeze()

        # pick components
        indexes = torch.multinomial(w, batch_size, replacement=
    True)

        # means and logvars
        eps = torch.randn(batch_size, self.L)
        for i in range(batch_size):
            indx = indexes[i]
            if i == 0:
                z = means[[indx]] + eps[[i]] * torch.exp(logvars
    [[indx]])
            else:
                z = torch.cat((z, means[[indx]] + eps[[i]] *
    torch.exp(logvars[[indx]])), 0)
        return z

    def log_prob(self, z):
        # mu, lof_var
```

Fig. 4.9 An example of the
MoG prior (contours) and the
samples from the aggregated
posterior (black dots)

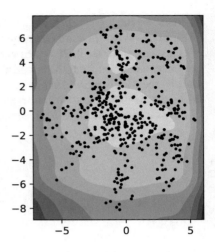

```
41        means, logvars = self.get_params()
42
43        # mixing probabilities
44        w = F.softmax(self.w, dim=0)
45
46        # log-mixture-of-Gaussians
47        z = z.unsqueeze(0) # 1 x B x L
48        means = means.unsqueeze(1) # K x 1 x L
49        logvars = logvars.unsqueeze(1) # K x 1 x L
50
51        log_p = log_normal_diag(z, means, logvars) + torch.log(w)
     # K x B x L
52        log_prob = torch.logsumexp(log_p, dim=0, keepdim=False) #
     B x L
53
54        return log_prob
```

Listing 4.7 A Mixture of Gaussians prior class

4.4.1.3 VampPrior: Variational Mixture of Posterior Prior

In [23], it was noticed that we can improve on the MoG prior and approximate the
aggregated posterior by introducing *pseudo-inputs*:

$$p_\lambda(\mathbf{z}) = \frac{1}{N} \sum_{k=1}^{K} q_\phi(\mathbf{z}|\mathbf{u}_k), \tag{4.48}$$

where $\lambda = \{\phi, \{\mathbf{u}_k^2\}\}$ are trainable parameters and $\mathbf{u}_k \in \mathcal{X}^D$ is a pseudo-input.
Notice that ϕ is a part of the trainable parameters. The idea of pseudo-input is to

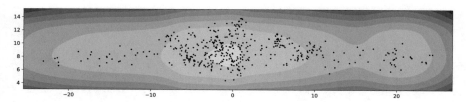

Fig. 4.10 An example of the VampPrior (contours) and the samples from the aggregated posterior (black dots)

randomly initialize objects that mimic observable variables (e.g., images) and learn them by backpropagation.

This approximation to the aggregated posterior is called the **variational mixture of posterior prior**, VampPrior for short. In [23] you can find some interesting properties and further analysis of the VampPrior. The main drawback of the VampPrior lies in initializing the pseudo-inputs; however, it serves as a good proxy to the aggregated posterior that improves the generative quality of the VAE, e.g., [10, 50, 51].

Alemi et al. [10] presented a nice connection of the VampPrior with the information-theoretic perspective on the VAE. They further proposed to introduce learnable probabilities of the components:

$$p_\lambda(\mathbf{z}) = \sum_{k=1}^{K} w_k q_\phi(\mathbf{z}|\mathbf{u}_k), \tag{4.49}$$

to allow the VampPrior to select more relevant components (i.e., pseudo-inputs).

As in the previous cases, we train a small VAE with the VampPrior (with $K = 16$) and a two-dimensional latent space. In Fig. 4.10, we present samples from the encoder for the test data (black dots) and the contour plot for the VampPrior. Similar to the MoG prior, the VampPrior fits better the aggregated posterior and has fewer holes. In this case, we can see that the VampPrior allows the encoders to spread across the latent space (notice the values).

An example of an implementation of the VampPrior is presented below:

```
class VampPrior(nn.Module):
    def __init__(self, L, D, num_vals, encoder, num_components,
    data=None):
        super(VampPrior, self).__init__()

        self.L = L
        self.D = D
        self.num_vals = num_vals

        self.encoder = encoder

        # pseudo-inputs
        u = torch.rand(num_components, D) * self.num_vals
```

```python
13        self.u = nn.Parameter(u)
14
15        # mixing weights
16        self.w = nn.Parameter(torch.zeros(self.u.shape[0], 1, 1))
   # K x 1 x 1
17
18    def get_params(self):
19        # u->encoder->mu, lof_var
20        mean_vampprior, logvar_vampprior = self.encoder.encode(
   self.u) #(K x L), (K x L)
21        return mean_vampprior, logvar_vampprior
22
23    def sample(self, batch_size):
24        # u->encoder->mu, lof_var
25        mean_vampprior, logvar_vampprior = self.get_params()
26
27        # mixing probabilities
28        w = F.softmax(self.w, dim=0) # K x 1 x 1
29        w = w.squeeze()
30
31        # pick components
32        indexes = torch.multinomial(w, batch_size, replacement=
   True)
33
34        # means and logvars
35        eps = torch.randn(batch_size, self.L)
36        for i in range(batch_size):
37            indx = indexes[i]
38            if i == 0:
39                z = mean_vampprior[[indx]] + eps[[i]] * torch.exp
   (logvar_vampprior[[indx]])
40            else:
41                z = torch.cat((z, mean_vampprior[[indx]] + eps[[i
   ]] * torch.exp(logvar_vampprior[[indx]])), 0)
42        return z
43
44    def log_prob(self, z):
45        # u->encoder->mu, lof_var
46        mean_vampprior, logvar_vampprior = self.get_params() # (K
   x L) & (K x L)
47
48        # mixing probabilities
49        w = F.softmax(self.w, dim=0) # K x 1 x 1
50
51        # log-mixture-of-Gaussians
52        z = z.unsqueeze(0) # 1 x B x L
53        mean_vampprior = mean_vampprior.unsqueeze(1) # K x 1 x L
54        logvar_vampprior = logvar_vampprior.unsqueeze(1) # K x 1
   x L
55
56        log_p = log_normal_diag(z, mean_vampprior,
   logvar_vampprior) + torch.log(w) # K x B x L
57        log_prob = torch.logsumexp(log_p, dim=0, keepdim=False) #
   B x L
```

58
```
59          return log_prob
```

Listing 4.8 A VampPrior class

4.4.1.4   GTM: Generative Topographic Mapping

In fact, we can use any density estimator to model the prior. In [52] a density estimator called **generative topographic mapping** (GTM) was proposed that defines a grid of K points in a low-dimensional space, $\mathbf{u} \in \mathbb{R}^C$, namely:

$$p(\mathbf{u}) = \sum_{k=1}^{K} w_k \delta(\mathbf{u} - \mathbf{u}_k) \tag{4.50}$$

that is further transformed to a higher-dimensional space by a transformation g_γ. The transformation g_γ predicts parameters of a distribution, e.g., the Gaussian distribution and, thus, $g_\gamma : \mathbb{R}^C \to \mathbb{R}^{2 \times M}$. Eventually, we can define the distribution as follows:

$$p_\lambda(\mathbf{z}) = \int p(\mathbf{u}) \mathcal{N}\left(\mathbf{z} | \mu_g(\mathbf{u}), \sigma_g^2(\mathbf{u})\right) d\mathbf{u} \tag{4.51}$$

$$= \sum_{k=1}^{K} w_k \mathcal{N}\left(\mathbf{z} | \mu_g(\mathbf{u}_k), \sigma_g^2(\mathbf{u}_k)\right), \tag{4.52}$$

where $\mu_g(\mathbf{u})$ and σ_g^2 rare outputs of the transformation $g_\gamma(\mathbf{u})$.

For instance, for $C = 2$ and $K = 3$, we can define the following grid: $\mathbf{u} \in \{[-1, -1], [-1, 0], [-1, 1], [0, -1], [0, 1], [0, 1], [1, -1], [1, 0], [1, -1]\}$. Notice that the grid is fixed and only the transformation (e.g., a neural network) g_γ is trained.

As in the previous cases, we train a small VAE with the GTM-based prior (with $K = 16$, i.e., a 4×4 grid) and a two-dimensional latent space. In Fig. 4.11, we present samples from the encoder for the test data (black dots) and the contour plot for the GTM-based prior. Similar to the MoG prior and the VampPrior, the GTM-based prior learns a pretty flexible distribution.

An example of an implementation of the GTM-based prior is presented below:

```
1  class GTMPrior(nn.Module):
2      def __init__(self, L, gtm_net, num_components, u_min=-1.,
       u_max=1.):
3          super(GTMPrior, self).__init__()
4
5          self.L = L
6
7          # 2D manifold
```

Fig. 4.11 An example of the
GTM-based prior (contours)
and the samples from the
aggregated posterior (black
dots)

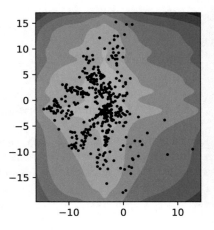

```
8     self.u = torch.zeros(num_components**2, 2) # K**2 x 2
9     u1 = torch.linspace(u_min, u_max, steps=num_components)
10    u2 = torch.linspace(u_min, u_max, steps=num_components)
11
12    k = 0
13    for i in range(num_components):
14        for j in range(num_components):
15            self.u[k,0] = u1[i]
16            self.u[k,1] = u2[j]
17            k = k + 1
18
19    # gtm network: u -> z
20    self.gtm_net = gtm_net
21
22    # mixing weights
23    self.w = nn.Parameter(torch.zeros(num_components**2, 1, 1))
24
25    def get_params(self):
26        # u->z
27        h_gtm = self.gtm_net(self.u) #K**2 x 2L
28        mean_gtm, logvar_gtm = torch.chunk(h_gtm, 2, dim=1) # K
      **2 x L and K**2 x L
29        return mean_gtm, logvar_gtm
30
31    def sample(self, batch_size):
32        # u->z
33        mean_gtm, logvar_gtm = self.get_params()
34
35        # mixing probabilities
36        w = F.softmax(self.w, dim=0)
37        w = w.squeeze()
38
39        # pick components
40        indexes = torch.multinomial(w, batch_size, replacement=
      True)
```

```
41
42        # means and logvars
43        eps = torch.randn(batch_size, self.L)
44        for i in range(batch_size):
45            indx = indexes[i]
46            if i == 0:
47                z = mean_gtm[[indx]] + eps[[i]] * torch.exp(
      logvar_gtm[[indx]])
48            else:
49                z = torch.cat((z, mean_gtm[[indx]] + eps[[i]] *
      torch.exp(logvar_gtm[[indx]])), 0)
50        return z
51
52    def log_prob(self, z):
53        # u->z
54        mean_gtm, logvar_gtm = self.get_params()
55
56        # log-mixture-of-Gaussians
57        z = z.unsqueeze(0) # 1 x B x L
58        mean_gtm = mean_gtm.unsqueeze(1) # K**2 x 1 x L
59        logvar_gtm = logvar_gtm.unsqueeze(1) # K**2 x 1 x L
60
61        w = F.softmax(self.w, dim=0)
62
63        log_p = log_normal_diag(z, mean_gtm, logvar_gtm) + torch.
      log(w) # K**2 x B x L
64        log_prob = torch.logsumexp(log_p, dim=0, keepdim=False) #
      B x L
65
66        return log_prob
```

Listing 4.9 A GTM-based prior class

4.4.1.5 GTM-VampPrior

As mentioned earlier, the main issue with the VampPrior is the initialization of the pseudo-inputs. Instead, we can use the idea of the GTM to learn the pseudo-inputs. Combining these two approaches, we get the following prior:

$$p_\lambda(\mathbf{z}) = \sum_{k=1}^{K} w_k q_\phi \left(\mathbf{z} | g_\gamma(\mathbf{u}_k) \right), \qquad (4.53)$$

where we first define a grid in a low-dimensional space, $\{\mathbf{u}_k\}$, and then transform them to $\mathcal{X}^D$ using the transformation g_γ.

Now, we train a small VAE with the GTM-VampPrior (with $K = 16$, i.e., a 4×4 grid) and a two-dimensional latent space. In Fig. 4.12, we present samples from the encoder for the test data (black dots) and the contour plot for the GTM-VampPrior.

Fig. 4.12 An example of the GTM-VampPrior (contours) and the samples from the aggregated posterior (black dots)

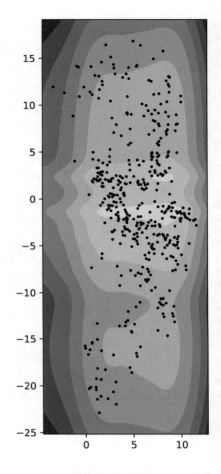

Again, this mixture-based prior allows to wrap the points (the aggregated posterior) and assign the probability to proper regions.

An example of an implementation of the GTM-VampPrior is presented below:

```
class GTMVampPrior(nn.Module):
    def __init__(self, L, D, gtm_net, encoder, num_points, u_min
    =-10., u_max=10., num_vals=255):
        super(GTMVampPrior, self).__init__()

        self.L = L
        self.D = D
        self.num_vals = num_vals

        self.encoder = encoder

        # 2D manifold
        self.u = torch.zeros(num_points**2, 2) # K**2 x 2
        u1 = torch.linspace(u_min, u_max, steps=num_points)
```

```python
u2 = torch.linspace(u_min, u_max, steps=num_points)

k = 0
for i in range(num_points):
    for j in range(num_points):
        self.u[k,0] = u1[i]
        self.u[k,1] = u2[j]
        k = k + 1

# gtm network: u -> x
self.gtm_net = gtm_net

# mixing weights
self.w = nn.Parameter(torch.zeros(num_points**2, 1, 1))

def get_params(self):
    # u->gtm_net->u_x
    h_gtm = self.gtm_net(self.u) #K x D
    h_gtm = h_gtm * self.num_vals
    # u_x->encoder->mu, lof_var
    mean_vampprior, logvar_vampprior = self.encoder.encode(
h_gtm) #(K x L), (K x L)
    return mean_vampprior, logvar_vampprior

def sample(self, batch_size):
# u->encoder->mu, lof_var
mean_vampprior, logvar_vampprior = self.get_params()

# mixing probabilities
w = F.softmax(self.w, dim=0)
w = w.squeeze()

    # pick components
    indexes = torch.multinomial(w, batch_size, replacement=
True)

    # means and logvars
    eps = torch.randn(batch_size, self.L)
    for i in range(batch_size):
        indx = indexes[i]
        if i == 0:
            z = mean_vampprior[[indx]] + eps[[i]] * torch.exp
(logvar_vampprior[[indx]])
        else:
            z = torch.cat((z, mean_vampprior[[indx]] + eps[[i
]] * torch.exp(logvar_vampprior[[indx]])), 0)
    return z

def log_prob(self, z):
    # u->encoder->mu, lof_var
    mean_vampprior, logvar_vampprior = self.get_params()

    # mixing probabilities
    w = F.softmax(self.w, dim=0)
```

```
64
65        # log-mixture-of-Gaussians
66        z = z.unsqueeze(0) # 1 x B x L
67        mean_vampprior = mean_vampprior.unsqueeze(1) # K x 1 x L
68        logvar_vampprior = logvar_vampprior.unsqueeze(1) # K x 1
      x L
69
70        log_p = log_normal_diag(z, mean_vampprior,
      logvar_vampprior) + torch.log(w) # K x B x L
71        log_prob = torch.logsumexp(log_p, dim=0, keepdim=False) #
      B x L
72
73        return log_prob
```

Listing 4.10 A GTM-VampPrior prior class

4.4.1.6 Flow-Based Prior

The last distribution we want to discuss here is a flow-based prior. Since flow-based models can be used to estimate any distribution, it is almost obvious to use them for approximating the aggregated posterior. Here, we use the implementation of the RealNVP presented before (see Chap. 3 for details).

As in the previous cases, we train a small VAE with the flow-based prior and two-dimensional latent space. In Fig. 4.13, we present samples from the encoder for the test data (black dots) and the contour plot for the flow-based prior. Similar to the previous mixture-based priors, the flow-based prior allows approximating the aggregated posterior very well. This is in line with many papers using flows as the prior in the VAE [24, 25]; however, we must remember that the flexibility of the flow-based prior comes with the cost of an increased number of parameters and potential training issues inherited from the flows.

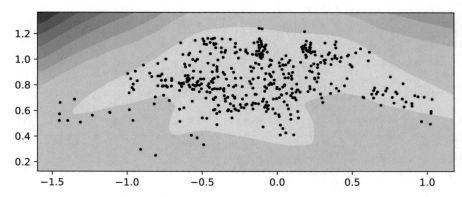

Fig. 4.13 An example of the flow-based prior (contours) and the samples from the aggregated posterior (black dots)

An example of an implementation of the flow-based prior is presented below:

```python
class FlowPrior(nn.Module):
    def __init__(self, nets, nett, num_flows, D=2):
        super(FlowPrior, self).__init__()

        self.D = D

        self.t = torch.nn.ModuleList([nett() for _ in range(
            num_flows)])
        self.s = torch.nn.ModuleList([nets() for _ in range(
            num_flows)])
        self.num_flows = num_flows

    def coupling(self, x, index, forward=True):
        (xa, xb) = torch.chunk(x, 2, 1)

        s = self.s[index](xa)
        t = self.t[index](xa)

        if forward:
            #yb = f^{-1}(x)
            yb = (xb - t) * torch.exp(-s)
        else:
            #xb = f(y)
            yb = torch.exp(s) * xb + t

        return torch.cat((xa, yb), 1), s

    def permute(self, x):
        return x.flip(1)

    def f(self, x):
        log_det_J, z = x.new_zeros(x.shape[0]), x
        for i in range(self.num_flows):
            z, s = self.coupling(z, i, forward=True)
            z = self.permute(z)
            log_det_J = log_det_J - s.sum(dim=1)

        return z, log_det_J

    def f_inv(self, z):
        x = z
        for i in reversed(range(self.num_flows)):
            x = self.permute(x)
            x, _ = self.coupling(x, i, forward=False)

        return x

    def sample(self, batch_size):
        z = torch.randn(batch_size, self.D)
        x = self.f_inv(z)
        return x.view(-1, self.D)
```

```
51    def log_prob(self, x):
52        z, log_det_J = self.f(x)
53        log_p = (log_standard_normal(z) + log_det_J.unsqueeze(1))
54        return log_p
```

Listing 4.11 A flow-based prior class

4.4.1.7   Remarks

In practice, we can use any density estimator to model $p_\lambda(\mathbf{z})$. For instance, we can use an autoregressive model [26] or more advanced approaches like resampled priors [27] or hierarchical priors [51]. Therefore, there are many options! However, there is still an open question **how** to do that and **what** role the prior (the marginal) should play. As I mentioned in the beginning, Bayesianists would say that the marginal should impose some constraints on the latent space or, in other words, our prior knowledge about it. I am a Bayesiast deep down in my heart and this way of thinking is very appealing to me. However, it is still unclear what is a good latent representation. This question is as old as mathematical modeling. I think that it would be interesting to look at optimization techniques, maybe applying a gradient-based method to all parameters/weights at once is not the best solution. Anyhow, I am pretty sure that modeling the prior is more important than many people think and plays a crucial role in VAEs.

4.4.2   Variational Posteriors

In general, variational inference searches for the best posterior approximation within a parametric family of distributions. Hence, recovering the true posterior is possible only if it happens to be in the chosen family. In particular, with widely used variational families such as diagonal covariance Gaussian distributions, the variational approximation is likely to be insufficient. Therefore, designing tractable and more expressive variational families is an important problem in VAEs. Here, we present two families of conditional normalizing flows that can be used for that purpose, namely, Householder flows [20] and Sylvester flows [16]. There are other interesting families and we refer the reader to the original papers, e.g., the generalized Sylvester flows [17] and the Inverse Autoregressive Flows [18].

The general idea about using the normalizing flows to parameterize the variational posteriors is to start with a relatively simple distribution like the Gaussian with the diagonal covariance matrix and then transform it to a complex distribution through a series of invertible transformations [19]. Formally speaking, we start with the latents $\mathbf{z}^{(0)}$ distributed according to $\mathcal{N}(\mathbf{z}^{(0)}|\mu(\mathbf{x}, \sigma^2(\mathbf{x}))$ and then after applying a series of invertible transformations $\mathbf{f}^{(t)}$, for $t = 1, \ldots, T$, the last iterate gives a random variable $\mathbf{z}^{(T)}$ that has a more flexible distribution. Once we choose transformations $\mathbf{f}^{(t)}$ for which the Jacobian-determinant can be computed, we aim at optimizing the following objective:

$$\ln p(\mathbf{x}) \geq \mathbb{E}_{q(\mathbf{z}^{(0)}|\mathbf{x})}\left[\ln p(\mathbf{x}|\mathbf{z}^{(T)}) + \sum_{t=1}^{T} \ln \left|\det \frac{\partial \mathbf{f}^{(t)}}{\partial \mathbf{z}^{(t-1)}}\right|\right] - \mathrm{KL}\big(q(\mathbf{z}^{(0)}|\mathbf{x})||p(\mathbf{z}^{(T)})\big).$$

(4.54)

In fact, the normalizing flow can be used to enrich the posterior of the VAE with small or even none modifications in the architecture of the encoder and the decoder.

4.4.2.1 Variational Posteriors with Householder Flows [20]

Motivation

First, we notice that any full-covariance matrix Σ can be represented by the eigenvalue decomposition using eigenvectors and eigenvalues:

$$\Sigma = \mathbf{U}\mathbf{D}\mathbf{U}^\top,$$

(4.55)

where $\mathbf{U}$ is an orthogonal matrix with eigenvectors in columns and $\mathbf{D}$ is a diagonal matrix with eigenvalues. In the case of the vanilla VAE, it would be tempting to model the matrix $\mathbf{U}$ to obtain a full-covariance matrix. The procedure would require a linear transformation of a random variable using an orthogonal matrix $\mathbf{U}$. Since the absolute value of the Jacobian-determinant of an orthogonal matrix is 1, for $\mathbf{z}^{(1)} = \mathbf{U}\mathbf{z}^{(0)}$ one gets $\mathbf{z}^{(1)} \sim \mathcal{N}(\mathbf{U}\boldsymbol{\mu}, \mathbf{U}\,\mathrm{diag}(\boldsymbol{\sigma}^2)\,\mathbf{U}^\top)$. If $\mathrm{diag}(\boldsymbol{\sigma}^2)$ coincides with true $\mathbf{D}$, then it would be possible to resemble the true full-covariance matrix. Hence, the main goal would be to model the orthogonal matrix of eigenvectors.

Generally, the task of modeling an orthogonal matrix in a principled manner is rather non-trivial. However, first we notice that any orthogonal matrix can be represented in the following form [53, 54]:

Theorem 4.1 (The Basis-Kernel Representation of Orthogonal Matrices) *For any $M \times M$ orthogonal matrix $\mathbf{U}$, there exist a full-rank $M \times K$ matrix $\mathbf{Y}$ (the basis) and a nonsingular (triangular) $K \times K$ matrix $\mathbf{S}$ (the kernel), $K \leq M$, such that:*

$$\mathbf{U} = \mathbf{I} - \mathbf{Y}\mathbf{S}\mathbf{Y}^\top.$$

(4.56)

The value K is called the *degree* of the orthogonal matrix. Further, it can be shown that any orthogonal matrix of degree K can be expressed using the product of Householder transformations [53, 54], namely:

Theorem 4.2 *Any orthogonal matrix with the basis acting on the K-dimensional subspace can be expressed as a product of exactly K Householder matrices:*

$$\mathbf{U} = \mathbf{H}_K \mathbf{H}_{K-1} \cdots \mathbf{H}_1,$$

(4.57)

where $\mathbf{H}_k = \mathbf{I} - \mathbf{S}_{kk}\mathbf{Y}_{.k}(\mathbf{Y}_{.k})^\top$, for $k = 1, \ldots, K$.

Theoretically, Theorem 4.2 shows that we can model any orthogonal matrix in a principled fashion using K Householder transformations. Moreover, the Householder matrix $\mathbf{H}_k$ is *orthogonal* matrix itself [55]. Therefore, this property and the Theorem 4.2 put the Householder transformation as a perfect candidate for formulating a volume-preserving flow that allows to approximate (or even capture) the true full-covariance matrix.

Householder Flows

The *Householder transformation* is defined as follows. For a given vector $\mathbf{z}^{(t-1)}$, the reflection hyperplane can be defined by a vector (a *Householder vector*) $\mathbf{v}_t \in \mathbb{R}^M$ that is orthogonal to the hyperplane, and the reflection of this point about the hyperplane is [55]

$$\mathbf{z}^{(t)} = \left(\mathbf{I} - 2\frac{\mathbf{v}_t\mathbf{v}_t^\top}{||\mathbf{v}_t||^2}\right)\mathbf{z}^{(t-1)} \tag{4.58}$$

$$= \mathbf{H}_t\mathbf{z}^{(t-1)}, \tag{4.59}$$

where $\mathbf{H}_t = \mathbf{I} - 2\frac{\mathbf{v}_t\mathbf{v}_t^\top}{||\mathbf{v}_t||^2}$ is called the *Householder matrix*.

The most important property of $\mathbf{H}_t$ is that it is an orthogonal matrix and hence the absolute value of the Jacobian-determinant is equal to 1. This fact significantly simplifies the objective (4.54) because $\ln\left|\det\frac{\partial \mathbf{H}_t\mathbf{z}^{(t-1)}}{\partial \mathbf{z}^{(t-1)}}\right| = 0$, for $t = 1, \ldots, T$. Starting from a simple posterior with the diagonal covariance matrix for $\mathbf{z}^{(0)}$, the series of T linear transformations given by (4.58) defines a new type of volume-preserving flow that we refer to as the *Householder flow* (HF). The vectors $\mathbf{v}_t$, $t = 1, \ldots, T$, are produced by the encoder network along with means and variances using a linear layer with the input $\mathbf{v}_{t-1}$, where $\mathbf{v}_0 = \mathbf{h}$ is the last hidden layer of the encoder network. The idea of the Householder flow is schematically presented in Fig. 4.14. Once the encoder returns the first Householder vector, the Householder

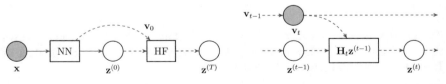

(a) encoder network + Householder Flow   (b) one step of the Householder Flow

Fig. 4.14 A schematic representation of the encoder network with the Householder flow. (**a**) The general architecture of the VAE+HF: The encoder returns means and variances for the posterior and the first Householder vector that is further used to formulate the Householder flow. (**b**) A single step of the Householder flow that uses linear Householder transformation. In both panels solid lines correspond to the encoder network and the dashed lines are additional quantities required by the HF

flow requires T linear operations to produce a sample from a more flexible posterior with an approximate full-covariance matrix.

4.4.2.2 Variational Posteriors with Sylvester Flows [16]

Motivation

The Householder flows can model only full-covariance Gaussians that is still not necessarily a rich family of distributions. Now, we will look into a generalization of the Householder flows. For this purpose, let us consider the following transformation similar to a single layer MLP with M hidden units and a residual connection:

$$\mathbf{z}^{(t)} = \mathbf{z}^{(t-1)} + \mathbf{A}h(\mathbf{B}\mathbf{z}^{(t-1)} + \mathbf{b}), \qquad (4.60)$$

with $\mathbf{A} \in \mathbb{R}^{D \times M}$, $\mathbf{B} \in \mathbb{R}^{M \times D}$, $\mathbf{b} \in \mathbb{R}^{M}$, and $M \leq D$. The Jacobian-determinant of this transformation can be obtained using *Sylvester's determinant identity*, which is a generalization of the matrix determinant lemma.

Theorem 4.3 (Sylvester's Determinant Identity) *For all* $\mathbf{A} \in \mathbb{R}^{D \times M}$, $\mathbf{B} \in \mathbb{R}^{M \times D}$,

$$\det(\mathbf{I}_D + \mathbf{A}\mathbf{B}) = \det(\mathbf{I}_M + \mathbf{B}\mathbf{A}), \qquad (4.61)$$

where $\mathbf{I}_M$ *and* $\mathbf{I}_D$ *are* M- *and* D-*dimensional identity matrices, respectively.*

When $M < D$, the computation of the determinant of a $D \times D$ matrix is thus reduced to the computation of the determinant of an $M \times M$ matrix.

Using Sylvester's determinant identity, the Jacobian-determinant of the transformation in Eq. (4.60) is given by:

$$\det\left(\frac{\partial \mathbf{z}^{(t)}}{\partial \mathbf{z}^{(t-1)}}\right) = \det\left(\mathbf{I}_M + \operatorname{diag}\left(h'(\mathbf{B}\mathbf{z}^{(t-1)} + \mathbf{b})\right)\mathbf{B}\mathbf{A}\right). \qquad (4.62)$$

Since Sylvester's determinant identity plays a crucial role in the proposed family of normalizing flows, we will refer to them as *Sylvester normalizing flows*.

Parameterization of $\mathbf{A}$ and $\mathbf{B}$

In general, the transformation in (4.60) will not be invertible. Therefore, we propose the following special case of the above transformation:

$$\mathbf{z}^{(t)} = \mathbf{z}^{(t-1)} + \mathbf{Q}\mathbf{R}h(\tilde{\mathbf{R}}\mathbf{Q}^T\mathbf{z}^{(t-1)} + \mathbf{b}), \qquad (4.63)$$

where $\mathbf{R}$ and $\tilde{\mathbf{R}}$ are upper triangular $M \times M$ matrices, and

$$\mathbf{Q} = (\mathbf{q}_1 \dots \mathbf{q}_M)$$

with the columns $\mathbf{q}_m \in \mathbb{R}^D$ forming an orthonormal set of vectors. By Theorem 4.3, the determinant of the Jacobian $\mathbf{J}$ of this transformation reduces to:

$$\det(\mathbf{J}) = \det\left(\mathbf{I}_M + \text{diag}\left(h'(\tilde{\mathbf{R}}\mathbf{Q}^T \mathbf{z}^{(t-1)} + \mathbf{b})\right)\tilde{\mathbf{R}}\mathbf{Q}^T \mathbf{Q}\mathbf{R}\right)$$

$$= \det\left(\mathbf{I}_M + \text{diag}\left(h'(\tilde{\mathbf{R}}\mathbf{Q}^T \mathbf{z}^{(t-1)} + \mathbf{b})\right)\tilde{\mathbf{R}}\mathbf{R}\right), \qquad (4.64)$$

which can be computed in $O(M)$, since $\tilde{\mathbf{R}}\mathbf{R}$ is also upper triangular. The following theorem gives a sufficient condition for this transformation to be invertible.

Theorem 4.4 *Let $\mathbf{R}$ and $\tilde{\mathbf{R}}$ be upper triangular matrices. Let $h : \mathbb{R} \longrightarrow \mathbb{R}$ be a smooth function with bounded, positive derivative. Then, if the diagonal entries of $\mathbf{R}$ and $\tilde{\mathbf{R}}$ satisfy $r_{ii}\tilde{r}_{ii} > -1/\|h'\|_\infty$ and $\tilde{\mathbf{R}}$ is invertible, the transformation given by (4.63) is invertible.*

The proof of this theorem could be found in [16].

Preserving Orthogonality of $\mathbf{Q}$

Orthogonality is a convenient property, mathematically, but hard to achieve in practice. In this chapter we consider three different flows based on the theorem above and various ways to preserve the orthogonality of $\mathbf{Q}$. The first two use explicit differentiable constructions of orthogonal matrices, while the third variant assumes a specific fixed permutation matrix as the orthogonal matrix.

Orthogonal Sylvester Flows First, we consider a Sylvester flow using matrices with M orthogonal columns (O-SNF). In this flow we can choose $M < D$ and thus introduce a flexible bottleneck. Similar to [56], we ensure orthogonality of $\mathbf{Q}$ by applying the following differentiable iterative procedure proposed by Björck and Bowie [57] and Kovarik [58]:

$$\mathbf{Q}^{(k+1)} = \mathbf{Q}^{(k)}\left(\mathbf{I} + \frac{1}{2}\left(\mathbf{I} - \mathbf{Q}^{(k)\top}\mathbf{Q}^{(k)}\right)\right), \qquad (4.65)$$

with a sufficient condition for convergence given by $\|\mathbf{Q}^{(0)\top}\mathbf{Q}^{(0)} - \mathbf{I}\|_2 < 1$. Here, the 2-norm of a matrix $\mathbf{X}$ refers to $\|\mathbf{X}\|_2 = \lambda_{\max}(\mathbf{X})$, with $\lambda_{\max}(\mathbf{X})$ representing the largest singular value of $\mathbf{X}$. In our experimental evaluations we ran the iterative procedure until $\|\mathbf{Q}^{(k)\top}\mathbf{Q}^{(k)} - \mathbf{I}\|_F \le \epsilon$, with $\|\mathbf{X}\|_F$ the Frobenius norm, and ϵ a small convergence threshold. We observed that running this procedure up to 30 steps was sufficient to ensure convergence with respect to this threshold. To minimize the computational overhead introduced by orthogonalization, we perform this orthogonalization in parallel for all flows.

Since this orthogonalization procedure is differentiable, it allows for the calculation of gradients with respect to $\mathbf{Q}^{(0)}$ by backpropagation, allowing for any standard optimization scheme such as stochastic gradient descent to be used for updating the flow parameters.

Householder Sylvester Flows Second, we study Householder Sylvester flows (H-SNF) where the orthogonal matrices are constructed by products of Householder reflections. Householder transformations are reflections about hyperplanes. Let $\mathbf{v} \in \mathbb{R}^D$, then the reflection about the hyperplane orthogonal to $\mathbf{v}$ is given by Eq. (4.58).

It is worth noting that performing a single Householder transformation is very cheap to compute, as it only requires D parameters. Chaining together several Householder transformations results in more general orthogonal matrices, and Theorem 4.2 shows that any $M \times M$ orthogonal matrix can be written as the product of $M - 1$ Householder transformations. In our Householder Sylvester flow, the number of Householder transformations H is a hyperparameter that trades off the number of parameters and the generality of the orthogonal transformation. Note that the use of Householder transformations forces us to use $M = D$, since Householder transformations result in square matrices.

Triangular Sylvester Flows Third, we consider a triangular Sylvester flow (T-SNF), in which all orthogonal matrices $\mathbf{Q}$ alternate per transformation between the identity matrix and the permutation matrix corresponding to reversing the order of $\mathbf{z}$. This is equivalent to alternating between lower and upper triangular $\tilde{\mathbf{R}}$ and $\mathbf{R}$ for each flow.

Amortizing Flow Parameters

When using normalizing flows in an amortized inference setting, the parameters of the base distribution as well as the flow parameters can be functions of the datapoint $\mathbf{x}$ [19]. Figure 4.15 (left) shows a diagram of one SNF step and the amortization procedure. The inference network takes datapoints $\mathbf{x}$ as input and provides as an output the mean and variance of $\mathbf{z}^{(0)}$ such that $\mathbf{z}^{(0)} \sim \mathcal{N}(\mathbf{z}|\mu^0, \sigma^0)$. Several SNF transformations are then applied to $\mathbf{z}^{(0)} \rightarrow \mathbf{z}^{(1)} \rightarrow \ldots \mathbf{z}^{(T)}$, producing a flexible posterior distribution for $\mathbf{z}^{(T)}$. All of the flow parameters ($\mathbf{R}$, $\tilde{\mathbf{R}}$, and $\mathbf{Q}$ for each transformation) are produced as an output by the inference network and are thus fully amortized.

4.4.2.3 Hyperspherical Latent Space

Motivation

In the VAE framework, choosing Gaussian priors and Gaussian posteriors from the mathematical convenience leads to Euclidean latent space. However, such choice could be limiting for the following reasons:

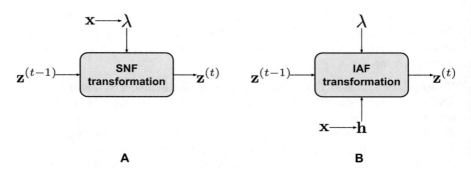

Fig. 4.15 Different amortization strategies for Sylvester normalizing flows and Inverse Autoregressive Flows. (**a**) Our inference network produces amortized flow parameters. This strategy is also employed by planar flows. (**b**) Inverse Autoregressive Flow [18] introduces a measure of **x** dependence through a context variable **h**(**x**). This context acts as an additional input for each transformation. The flow parameters themselves are independent of **x**

- In low dimensions, the standard Gaussian probability presents a concentrated probability mass around the mean, encouraging points to cluster in the center. However, this is particularly problematic when the data is divided into multiple clusters. Then, a better suited prior would be *uniform*. Such a uniform prior, however, is not well-defined on the hyperplane.
- It is a well-known phenomenon that the standard Gaussian distribution in high dimensions tends to resemble a uniform distribution on the surface of a hypersphere, with the vast majority of its mass concentrated on the hyperspherical shell (the so-called soap bubble effect). A natural question is whether it would be better to use a distribution defined on the hypersphere.

A distribution that would allow solving both problems at once is the von Mises–Fisher distribution. It was advocated in [33] to use this distribution in the context of VAEs.

von Mises–Fisher Distribution

The *von Mises–Fisher* (vMF) distribution is often described as the Normal Gaussian distribution on a hypersphere. Analogous to a Gaussian, it is parameterized by $\mu \in \mathbb{R}^m$ indicating the mean direction, and $\kappa \in \mathbb{R}_{\geq 0}$ the concentration around μ. For the special case of $\kappa = 0$, the vMF represents a Uniform distribution. The probability density function of the vMF distribution for a random unit vector $\mathbf{z} \in \mathbb{R}^m$ (or $\mathbf{z} \in \mathcal{S}^{m-1}$) is then defined as

$$q(\mathbf{z}|\boldsymbol{\mu}, \kappa) = C_m(\kappa) \exp\left(\kappa \boldsymbol{\mu}^T \mathbf{z}\right) \tag{4.66}$$

$$C_m(\kappa) = \frac{\kappa^{m/2-1}}{(2\pi)^{m/2} \mathcal{I}_{m/2-1}(\kappa)}, \tag{4.67}$$

where $||\boldsymbol{\mu}||^2 = 1$, $C_m(\kappa)$ is the normalizing constant, and $\mathcal{I}_v$ denotes the modified Bessel function of the first kind at order v.

Interestingly, since we define a distribution over a hypersphere, it is possible to formulate a uniform prior over the hypersphere. Then it turns out that if we take the vMF distribution as the variational posterior, it is possible to calculate the Kullback–Leibler divergence between the vMF distribution and the uniform defined over $\mathcal{S}^{m-1}$ analytically [33]:

$$KL[\text{vMF}(\mu, \kappa)||\text{Unif}(\mathcal{S}^{m-1})] = \kappa + \log C_m(\kappa) - \log\left(\frac{2(\pi^{m/2})}{\Gamma(m/2)}\right)^{-1}. \tag{4.68}$$

To sample from the vMF, one can follow the procedure of [59]. Importantly, the reparameterization cannot be easily formulated for the vMF distribution. Fortunately, [60] allows extending the reparameterization trick to the wide class of distributions that can be simulated using rejection sampling. [33] presents how to formulate the acceptance–rejection sampling reparameterization procedure. Being equipped with the sampling procedure and the reparameterization trick, and having an analytical form of the Kullback–Leibler divergence, we have everything to be able to build a hyperspherical VAE. However, please note the all these procedures are less trivial than the ones for Gaussians. Therefore, a curious reader is referred to [33] for further details.

4.5   Hierarchical Latent Variable Models

4.5.1   Introduction

The main goal of AI is to formulate and implement systems that can interact with an environment, process, store, and transmit information. In other words, we wish an AI system *understands* the world around it by identifying and disentangling hidden factors in the observed low-sensory data [61]. If we think about the problem of building such a system, we can formulate it as learning a probabilistic model, i.e., a joint distribution over observed data, $\mathbf{x}$, and hidden factors, $\mathbf{z}$, namely, $p(\mathbf{x}, \mathbf{z})$. Then learning a *useful representation* is equivalent to finding a posterior distribution over the hidden factors, $p(\mathbf{z}|\mathbf{x})$. However, it is rather unclear what we really mean by *useful* in this context. In a beautiful blog post [62], Ferenc Huszar outlines why learning a latent variable model by maximizing the likelihood function is not necessarily useful from the representation learning perspective. Here, we will use it as a good starting point for a discussion of why applying hierarchical latent variable models could be beneficial.

Let us start by defining the setup. We assume the empirical distribution $p_{data}(\mathbf{x})$ and a latent variable model $p_\theta(\mathbf{x}, \mathbf{z})$. The way we parameterize the latent variable model is not constrained in any manner; however, we assume that the distribution is parameterized using deep neural networks (DNNs). This is important for two reasons:

1. DNNs are non-linear transformations and as such, they are flexible and allow parameterizing a wide range of distributions.
2. We must remember that DNNs **will not** solve all problems for us! In the end, we need to think about the model as a whole, not only about the parameterization. What I mean by that is the distribution we choose and how random variables interact, etc. DNNs are definitely helpful, but there are many potential pitfalls (we will discuss some of them later on) that even the largest and coolest DNN is unable to take care of.

It is worth to remember that the joint distribution could be factorized in two ways, namely:

$$p_\theta(\mathbf{x}, \mathbf{z}) = p_\theta(\mathbf{x}|\mathbf{z}) p_\theta(\mathbf{z}) \tag{4.69}$$

$$= p_\theta(\mathbf{z}|\mathbf{x}) p_\theta(\mathbf{x}). \tag{4.70}$$

Moreover, the training problem of learning θ could be defined as an unconstrained optimization problem with the following training objective:

$$KL[p_{data}(\mathbf{x})||p_\theta(\mathbf{x})] = -\mathbb{H}[p_{data}(\mathbf{x})] + \mathbb{CE}[p_{data}(\mathbf{x})||p_\theta(\mathbf{x})] \tag{4.71}$$

$$= const + \mathbb{CE}[p_{data}(\mathbf{x})||p_\theta(\mathbf{x})], \tag{4.72}$$

where $p_\theta(\mathbf{x}) = \int p_\theta(\mathbf{x}, \mathbf{z}) \, d\mathbf{z}$, $\mathbb{H}[\cdot]$ denotes the entropy, and $\mathbb{CE}[\cdot||\cdot]$ is the cross-entropy. Notice that the entropy of the empirical distribution is simply a constant since it does not contain θ. The cross-entropy could be further re-written as follows:

$$\mathbb{CE}[p_{data}(\mathbf{x})||p_\theta(\mathbf{x})] = -\int p_{data}(\mathbf{x}) \ln p_\theta(\mathbf{x}) \, d\mathbf{x} \tag{4.73}$$

$$= -\frac{1}{N} \sum_{n=1}^{N} \ln p_\theta(\mathbf{x}_n). \tag{4.74}$$

Eventually, we have obtained the objective function we use all the time, namely, the negative log-likelihood function.

If we think of *usefulness* of a representation (i.e., hidden factors) $\mathbf{z}$, we intuitively think of some kind of information that is shared between $\mathbf{z}$ and $\mathbf{x}$. However, the unconstrained training problem we consider, i.e., the minimization of the negative log-likelihood function, does not necessarily say **anything** about the latent

Fig. 4.16 A schematic diagram representing a dependency between *usefulness* and the objective function for all possible latent variable models. The darker the color, the better the objective function value. Reproduced based on [62]

All possible latent variable models

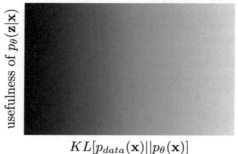

$$KL[p_{data}(\mathbf{x})||p_\theta(\mathbf{x})]$$

representation. In the end, we optimize the **marginal** over observable variables because we do not have access to values of latent variables. Even more, typically we do not know what these hidden factors are or should be! As a result, our latent variable model can learn to disregard the latent variables completely. Let us look into this problem in more detail.

A Potential Problem with Latent Variable Models

Following the discussion presented in [62], we can visualize two scenarios that are pretty common in deep generative modeling with latent variable models. Before delving into that, it is beneficial to explain the general picture. We are interested in analyzing a class of latent variable models with respect to *usefulness* of latents and the value of the objective function $KL[p_{data}(\mathbf{x})||p_\theta(\mathbf{x})]$. In Fig. 4.16, we depict a case when all models are possible, namely, a search space where models are evaluated according to the training objective (x-axis) and *usefulness* (y-axis). The ideal model is the one in the top-left corner that maximizes both criteria. However, it is possible to find a model that completely disregards the latents (the bottom-left corner) while maximizing the fit to data. We already can see that there is a potentially huge problem! Running a (numerical) optimization procedure could give infinitely many models that are equally good with respect to $KL[p_{data}(\mathbf{x})||p_\theta(\mathbf{x})]$ but with completely different posteriors over latents! That puts in question the applicability of the latent variable models. However, in practice, we see that learned latent variables are useful (or, in other words, they contain information about observables). So how is it possible?

As pointed out by Huszár [62], the reason for that is the inductive bias of the chosen class of models. By picking a very specific class of DNNs, we implicitly constrain the search space. First, the left-most models in Fig. 4.16 are typically unattainable. However, using some kind of bottlenecks in our class of models potentially leads to a situation that latents must contain some information about observables. As a result, they become *useful*. An example of such a situation is depicted in Fig. 4.17. After running a training algorithm, we can end in one of the two "spikes" where the training objective is the highest and the *usefulness* is non-

Fig. 4.17 A schematic diagram representing a dependency between *usefulness* and the objective function for a constrained class of models. The darker the color, the better the objective function value. Reproduced based on [62]

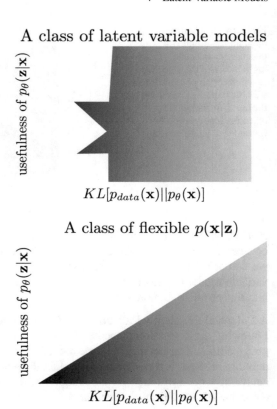

A class of latent variable models

usefulness of $p_\theta(\mathbf{z}|\mathbf{x})$

$KL[p_{data}(\mathbf{x})||p_\theta(\mathbf{x})]$

Fig. 4.18 A schematic diagram representing a dependency between *usefulness* and the objective function for a class of models with flexible $p(\mathbf{x}|\mathbf{z})$. The darker the color, the better the objective function value. Reproduced based on [62]

A class of flexible $p(\mathbf{x}|\mathbf{z})$

usefulness of $p_\theta(\mathbf{z}|\mathbf{x})$

$KL[p_{data}(\mathbf{x})||p_\theta(\mathbf{x})]$

zero. Still, we can achieve the same performing models at two different levels of the *usefulness* but at least the information flows from $\mathbf{x}$ to $\mathbf{z}$. Obviously, the considered scenario is purely hypothetical, but it shows that the inductive bias of a model can greatly help to learn representations without being specified by the objective function. Please keep this thought in mind because it will play a crucial role later on!

The next situation is more tricky. Let us assume that we have a constrained class of models; however, the conditional likelihood $p(\mathbf{x}|\mathbf{z})$ is parameterized by a flexible, enormous DNN. A potential danger here is that this model could learn to completely disregard $\mathbf{z}$, treating it as a noise. As a result, $p(\mathbf{x}|\mathbf{z})$ becomes an unconditional distribution that mimics $p_{data}(\mathbf{x})$ almost perfectly. At the first glance, this scenario sounds unrealistic, but it is a well-known phenomenon in the field. For instance, [10] conducted a thorough experiment with variational auto-encoders, and taking a PixelCNN++-based decoder resulted in a VAE that was unable to reconstruct images. Their conclusion was exactly the same, namely, taking a class of models with too flexible $p(\mathbf{x}|\mathbf{z})$ could lead to the model in the bottom-left corner in Fig. 4.18.

How to Define a *Proper* Class of Models?

Alright, you are probably a bit confused about what we have discussed so far. The general picture is rather pessimistic because it seems that picking a proper class of models, i.e., a class of models that allow achieving *useful* latent representations is a non-trivial task. Moreover, the whole story sounds like walking in the dark, trying out various DNN architectures, and hoping that we obtain a meaningful representation.

Fortunately, the problem is not so horrible as it looks at the first glance. Some ideas formulate a constrained optimization problem [12, 63] or add an auxiliary regularizer [64, 65] to (implicitly) define *usefulness* of the latents. Here, we will discuss one of the possible approaches that utilizes hierarchical architectures. However, it is worth remembering that the issue of learning *useful* representations remains an open question and is a vivid research direction.

Hierarchical models have a long history in deep generative modeling and deep learning and were advocated by many prominent researchers, e.g., [66–68]. The main hypothesis is that the concepts describing the world around us could be organized hierarchically. In the light of our discussion, if a latent variable model takes a hierarchical structure, it may introduce an inductive bias, constrain the class of models, and, eventually, force information flow between latents and observables. At least in theory. Shortly, we will see that we must be very careful with formulating stochastic dependencies in the hierarchy. In the next sections, we will focus on latent variable models with variational inference, i.e., hierarchical Variational Auto-Encoders.

A side note: One may be tempted to associate hierarchical modeling with Bayesian hierarchical modeling. These two terms are not necessarily equivalent. Bayesian hierarchical modeling is about treating (hyper)parameters as random variables and formulating distributions over (hyper)parameters [69]. Here, we do not take advantage of Bayesian modeling and consider a hierarchy among latent variables, not parameters.

4.5.2 Hierarchical VAEs

4.5.2.1 Two-Level VAEs

Let us start with a VAE with two latent variables: z_1 and z_2. The joint distribution could be factorized as follows:

$$p(\mathbf{x}, \mathbf{z}_1, \mathbf{z}_2) = p(\mathbf{x}|\mathbf{z}_1)p(\mathbf{z}_1|\mathbf{z}_2)p(\mathbf{z}_2). \tag{4.75}$$

This model defines a straightforward generative process: First sample z_2, then sample z_1 given z_2, and eventually sample $\mathbf{x}$ given z_1.

Fig. 4.19 An example of a
two-level VAE. (**a**) The
generative part. (**b**) The
variational part

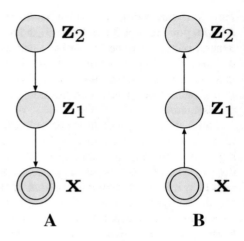

Since we know already that even for a single latent variable calculating posteriors
over latents is intractable (except the linear Gaussian case, it is worth remembering
that!), we can utilize the variational inference with a family of variational posteriors
$Q(\mathbf{z}_1, \mathbf{z}_2|\mathbf{x})$. Now, the main part is how to define the variational posteriors. A rather
natural approach would be to reverse the dependencies and factorize the posterior in
the following fashion:

$$Q(\mathbf{z}_1, \mathbf{z}_2|\mathbf{x}) = q(\mathbf{z}_1|\mathbf{x})q(\mathbf{z}_2|\mathbf{z}_1, \mathbf{x}), \tag{4.76}$$

or even we can simplify it as follows (dropping the dependency on $\mathbf{x}$ for the second
latent variable):

$$Q(\mathbf{z}_1, \mathbf{z}_2|\mathbf{x}) = q(\mathbf{z}_1|\mathbf{x})q(\mathbf{z}_2|\mathbf{z}_1). \tag{4.77}$$

If we take the continuous latents, we can use the Gaussian distributions:

$$p(\mathbf{z}_1|\mathbf{z}_2) = \mathcal{N}(\mathbf{z}_1|\mu(\mathbf{z}_2), \sigma^2(\mathbf{z}_2)) \tag{4.78}$$

$$p(\mathbf{z}_2) = \mathcal{N}(\mathbf{z}_2|0, 1) \tag{4.79}$$

$$q(\mathbf{z}_1|\mathbf{x}) = \mathcal{N}(\mathbf{z}_1|\mu(\mathbf{x}), \sigma^2(\mathbf{x})) \tag{4.80}$$

$$q(\mathbf{z}_2|\mathbf{z}_1) = \mathcal{N}(\mathbf{z}_2|\mu(\mathbf{z}_1), \sigma^2(\mathbf{z}_1)), \tag{4.81}$$

where $\mu_i(\mathbf{v})$ means that a mean parameter is parameterized by a neural network
that takes a random variable $\mathbf{v}$ as input, analogously we parameterize variances (i.e.,
diagonal covariance matrices). As we can see, this is a straightforward extension of
a VAE we discussed before.

The two-level VAE is depicted in Fig. 4.19. Notice how the stochastic dependencies are defined, namely, there is always a dependency on a single random variable.

A Potential Pitfall

Alright, so are we done? Do we have a better class of VAEs? Unfortunately, the answer is **no**. We noticed that this two-level version of a VAE is a rather straightforward extension of a one-level VAE. Thus, our discussion about potential problems with latent variable models holds true. We get even get an extra insight if we look into the ELBO for the two-level VAE (if you do not remember how to derive the ELBO, please go back to the post on VAEs first):

$$ELBO(\mathbf{x}) = \mathbb{E}_{Q(\mathbf{z}_1, \mathbf{z}_2|\mathbf{x})}\left[\ln p(\mathbf{x}|\mathbf{z}_1) - KL[q(\mathbf{z}_1|\mathbf{x})||p(\mathbf{z}_1|\mathbf{z}_2)] - KL[q(\mathbf{z}_2|\mathbf{z}_1)||p(\mathbf{z}_2)]\right]. \qquad (4.82)$$

To shed some light on the ELBO for the two-level VAE, we notice the following:

1. All conditions $(\mathbf{z}_1, \mathbf{z}_2, \mathbf{x})$ are either samples from $Q(\mathbf{z}_1, \mathbf{z}_2|\mathbf{x})$ or $p_{data}(\mathbf{x})$.
2. We obtain the Kullback–Leibler divergence terms by looking at the variables *per layer*. You are encouraged to derive the ELBO step-by-step, it is a great exercise to get familiar with the variational inference.
3. It is worth remembering that the Kullback–Leibler divergence is always non-negative.

Theoretically, everything should work perfectly fine, but there are a couple of potential problems. First, we initialize all DNNs that parameterize the distributions randomly. As a result, all Gaussians are basically standard Gaussians. Second, if the decoder is powerful and flexible, there is a huge danger that the model will try take advantage of the optimum for the last KL-term, $KL[q(\mathbf{z}_2|\mathbf{z}_1)||p(\mathbf{z}_2)]]$, that is, $q(\mathbf{z}_2|\mathbf{z}_1) \approx p(\mathbf{z}_2) \approx \mathcal{N}(0, 1)$. Then, since $q(\mathbf{z}_2|\mathbf{z}_1) \approx \mathcal{N}(0, 1)$, the second layer is not used at all (it is a Gaussian noise) and we get back to the same issues as in the one-level VAE architecture. It turns out that learning the two-level VAE is even more problematic than a VAE with a single latent because even for a relatively simple decoder the second latent variable $\mathbf{z}_2$ is mostly unused [15, 70]. This effect is called the *posterior collapse*.

4.5.2.2   Top-Down VAEs

A take-away from our considerations in the two-level VAE is that adding an extra level does not necessarily provide anything comparing to the one-level VAE. However, so far we have considered only one class of variational posteriors, namely:

$$Q(\mathbf{z}_1, \mathbf{z}_2|\mathbf{x}) = q(\mathbf{z}_1|\mathbf{x})q(\mathbf{z}_2|\mathbf{z}_1). \qquad (4.83)$$

A natural question is whether we can do better. You can already guess the answer, but before shouting it out loud, let us think for a second. In the generative part, we have *top-down* dependencies, going from the highest level of abstraction (latents) down to the observable variables. Let us repeat it here again:

$$p(\mathbf{x}, \mathbf{z}_1, \mathbf{z}_2) = p(\mathbf{x}|\mathbf{z}_1)p(\mathbf{z}_1|\mathbf{z}_2)p(\mathbf{z}_2). \qquad (4.84)$$

Perhaps, we can mirror such dependencies in the variational posteriors as well. Then we get the following:

$$Q(\mathbf{z}_1, \mathbf{z}_2|\mathbf{x}) = q(\mathbf{z}_1|\mathbf{z}_2, \mathbf{x})q(\mathbf{z}_2|\mathbf{x}). \qquad (4.85)$$

Do you see any resemblance? Yes, the variational posteriors have the extra **x**, but the dependencies are pointing in the same direction. Why this could be beneficial? Because now we could have a shared *top-down* path that would make the variational posteriors and the generative part tightly connected through a shared parameterization. That could be a very useful inductive bias!

This idea was originally proposed in ResNet VAEs [18] and Ladder VAEs [71], and it was further developed in BIVA [44], NVAE [45], and the very deep VAE [46]. These approaches differ in their implementations and parameterizations used (i.e., architectures of DNNs); however, they all could be categorized as instantiations of top-down VAEs. The main idea, as mentioned before, is to share the top-down path between the variational posteriors and the generative distributions and use a *side*, deterministic path going from **x** to the last latents. Alright, let us write this idea down.

First, we have the top-down path that defines $p(\mathbf{x}|\mathbf{z}_1)$, $p(\mathbf{z}_1|\mathbf{z}_2)$, and $p(\mathbf{z}_2)$. Thus, we need a DNN that outputs μ_1 and σ_1^2 for given $\mathbf{z}_2$, and another DNN that outputs the parameters of $p(\mathbf{x}|\mathbf{z}_1)$ for given $\mathbf{z}_1$. Since $p(\mathbf{z}_2)$ is an unconditional distribution (e.g., the standard Gaussian), we do not need a separate DNN here.

Second, we have a side, deterministic path that gives two deterministic variables: $\mathbf{r}_1 = f_1(\mathbf{x})$ and $\mathbf{r}_2 = f_2(\mathbf{r}_1)$. Both transformations, f_1 and f_2, are DNNs. Then, we can use additional DNNs that return some modifications of the means and the variances, namely, $\Delta\mu_1, \Delta\sigma_1^2$, and $\Delta\mu_2, \Delta\sigma_2^2$. These modifications could be defined in many ways. Here we follow the way it is done in NVAE [45], namely, the modifications are relative location and scales of the values given in the top-down path. If you do not fully follow this idea, it should be clear once we define the variational posteriors.

Finally, we can define the whole procedure. We define various neural networks by specifying different indices. For sampling, we use the top-down path:

1. $\mathbf{z}_2 \sim \mathcal{N}(0, 1)$
2. $[\mu_1, \sigma_1^2] = NN_1(\mathbf{z}_2)$
3. $\mathbf{z}_1 \sim \mathcal{N}(\mu_1, \sigma_1^2)$
4. $\vartheta = NN_x(\mathbf{z}_1)$
5. $\mathbf{x} \sim p_\vartheta(\mathbf{x}|\mathbf{z}_1)$

Now (please focus!) we calculate samples from the variational posteriors as follows:

Fig. 4.20 An example of the top-down VAE. Red nodes denote the deterministic path and blue nodes depict random variables

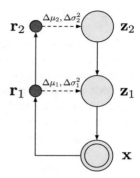

1. (*Bottom-up deterministic path*) $\mathbf{r}_1 = f_1(\mathbf{x})$ and $\mathbf{r}_2 = f_2(\mathbf{r}_1)$
2. $[\Delta\mu_1, \Delta\sigma_1^2] = NN_{\Delta 1}(r_1)$
3. $[\Delta\mu_2, \Delta\sigma_2^2] = NN_{\Delta 2}(r_2)$
4. $\mathbf{z}_2 \sim \mathcal{N}(0 + \Delta\mu_2, 1 \cdot \Delta\sigma_2^2)$
5. $[\mu_1, \sigma_1^2] = NN_1(\mathbf{z}_2)$
6. $\mathbf{z}_1 \sim \mathcal{N}(\mu_1 + \Delta\mu_1, \sigma_1^2 \cdot \Delta\sigma_1^2)$
7. $\vartheta = NN_x(\mathbf{z}_1)$
8. $\mathbf{x} \sim p_\vartheta(\mathbf{x}|\mathbf{z}_1)$

These operations are schematically presented in Fig. 4.20.

Please note that the deterministic bottom-up path modifies parameters of the top-down path. As advocated by Vahdat and Kautz [45], this idea is especially useful because "when the prior moves, the approximate posterior moves accordingly, if not changed." Moreover, as noted in [45], the Kullback–Leibler between two Gaussians simplifies as follows (we remove some additional dependencies for clarity):

$$KL\left(q\left(z_i \mid \boldsymbol{x}\right) \| p\left(z_i\right)\right) = \frac{1}{2}\left(\frac{\Delta\mu_i^2}{\sigma_i^2} + \Delta\sigma_i^2 - \log \Delta\sigma_i^2 - 1\right).$$

Eventually, we implicitly force a close connection between the variational posteriors and the generative part. This inductive bias helps to encode information about the observables in the latents. Moreover, there is no need to use overly flexible decoders since the latents take care of distilling the essence from data. I know, it is still a bit hand-wavy since we do not define the magical *usefulness*, but I hope you get the picture. The top-down VAEs entangle the variational posteriors and the generative path and, as a result, the Kullback–Leibler terms will not collapse (i.e., they will be greater than zero). Empirical studies strongly back up this hypothesis [44–46, 71].

4.5.2.3 Code

Let us delve into an implementation of a top-down VAE. We stick to the two-level VAE to match the description provided above. We will use precisely the same steps as in the procedures used above. For clarity, we will use a single class to the code as similar to the mathematical expressions above as possible. We use the reparameterization trick for sampling. There is one difference between the math and the code, namely, in the code we use $\log \Delta\sigma$ instead of $\Delta\sigma$. Then, we use $\log \sigma + \log \Delta\sigma$ instead of $\sigma \cdot \Delta\sigma$ because $e^{\log a + \log b} = e^{\log a} \cdot e^{\log b} = a \cdot b$.

```python
class HierarchicalVAE(nn.Module):
    def __init__(self, nn_r_1, nn_r_2, nn_delta_1, nn_delta_2,
    nn_z_1, nn_x, num_vals=256, D=64, L=16, likelihood_type='
    categorical'):
        super(HierarchicalVAE, self).__init__()

        print('Hierachical VAE by JT.')

        # bottom-up path
        self.nn_r_1 = nn_r_1
        self.nn_r_2 = nn_r_2

        self.nn_delta_1 = nn_delta_1
        self.nn_delta_2 = nn_delta_2

        # top-down path
        self.nn_z_1 = nn_z_1
        self.nn_x = nn_x


        # other params
        self.D = D # dim of inputs

        self.L = L # dim of the second latent layer

        self.num_vals = num_vals # num of values per pixel

        self.likelihood_type = likelihood_type # the conditional
    likelihood type (categorical/bernoulli)

    # If you don't remember the reparameterization trick, please
    go back to the post on VAEs.
    def reparameterization(self, mu, log_var):
        std = torch.exp(0.5*log_var)
        eps = torch.randn_like(std)
        return mu + std * eps

    def forward(self, x, reduction='avg'):
        #=====
        # First, we need to calculate the bottom-up deterministic
    path.
        # Here we use a small trick to keep the delta of variance
    contained, namely, we apply the hard-tanh non-linearity.
```

```
39      # bottom—up
40      # step 1
41      r_1 = self.nn_r_1(x)
42      r_2 = self.nn_r_2(r_1)
43
44      #step 2
45      delta_1 = self.nn_delta_1(r_1)
46      delta_mu_1, delta_log_var_1 = torch.chunk(delta_1, 2, dim
     =1)
47      delta_log_var_1 = F.hardtanh(delta_log_var_1, -7., 2.)
48
49      # step 3
50      delta_2 = self.nn_delta_2(r_2)
51      delta_mu_2, delta_log_var_2 = torch.chunk(delta_2, 2, dim
     =1)
52      delta_log_var_2 = F.hardtanh(delta_log_var_2, -7., 2.)
53
54      # Next, we can do the top—down path.
55
56      # top—down
57      # step 4
58      z_2 = self.reparameterization(delta_mu_2, delta_log_var_2
     )
59
60      # step 5
61      h_1 = self.nn_z_1(z_2)
62      mu_1, log_var_1 = torch.chunk(h_1, 2, dim=1)
63
64      # step 6
65      z_1 = self.reparameterization(mu_1 + delta_mu_1,
     log_var_1 + delta_log_var_1)
66
67      # step 7
68      h_d = self.nn_x(z_1)
69
70      if self.likelihood_type == 'categorical':
71          b = h_d.shape[0]
72          d = h_d.shape[1]//self.num_vals
73          h_d = h_d.view(b, d, self.num_vals)
74          mu_d = torch.softmax(h_d, 2)
75
76      elif self.likelihood_type == 'bernoulli':
77          mu_d = torch.sigmoid(h_d)
78
79      #=====ELBO
80      # RE
81      if self.likelihood_type == 'categorical':
82          RE = log_categorical(x, mu_d, num_classes=self.
     num_vals, reduction='sum', dim=-1).sum(-1)
83
84      elif self.likelihood_type == 'bernoulli':
85          RE = log_bernoulli(x, mu_d, reduction='sum', dim=-1)
86
87      # KL
```

```
88      # For the Kullback—Leibler part, we need calculate two
     divergences:
89      # 1) KL[q(z_2|z) || p(z_2)] where p(z_2) = N(0,1)
90      # 2) KL[q(z_1|z_2, x) || p(z_1|z_2)]
91      # Note: We use the analytical form of the KL between two
     Gaussians here. If you use a different distribution,
92      # please pay attention! You would need to use a different
      expression here.
93      KL_z_2 = 0.5 * (delta_mu_2**2 + torch.exp(delta_log_var_2
     ) - delta_log_var_2 - 1).sum(-1)
94      KL_z_1 = 0.5 * (delta_mu_1**2 / torch.exp(log_var_1) +
     torch.exp(delta_log_var_1) -\
95                      delta_log_var_1 - 1).sum(-1)
96
97      KL = KL_z_1 + KL_z_2
98
99      # Final ELBO
100     if reduction == 'sum':
101         loss = -(RE - KL).sum()
102     else:
103         loss = -(RE - KL).mean()
104
105     return loss
106
107 # Sampling is the top-down path but without calculating delta
     mean and delta variance.
108 def sample(self, batch_size=64):
109     # step 1
110     z_2 = torch.randn(batch_size, self.L)
111     # step 2
112     h_1 = self.nn_z_1(z_2)
113     mu_1, log_var_1 = torch.chunk(h_1, 2, dim=1)
114     # step 3
115     z_1 = self.reparameterization(mu_1, log_var_1)
116
117     # step 4
118     h_d = self.nn_x(z_1)
119
120     if self.likelihood_type == 'categorical':
121         b = batch_size
122         d = h_d.shape[1]//self.num_vals
123         h_d = h_d.view(b, d, self.num_vals)
124         mu_d = torch.softmax(h_d, 2)
125         # step 5
126         p = mu_d.view(-1, self.num_vals)
127         x_new = torch.multinomial(p, num_samples=1).view(b,d)
128
129     elif self.likelihood_type == 'bernoulli':
130         mu_d = torch.sigmoid(h_d)
131         # step 5
132         x_new = torch.bernoulli(mu_d)
133     return x_new
```

Listing 4.12 A top-down VAE class

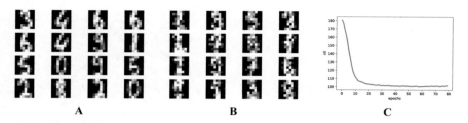

Fig. 4.21 An example of outcomes after the training: (**a**) Randomly selected real images. (**b**) Unconditional generations from the top-down VAE. (**c**) The validation curve during training

That's it! Now we are ready to run the full code (take a look at: https://github.com/jmtomczak/intro_dgm). After training our top-down VAE, we should obtain results like in Fig. 4.21.

4.5.2.4 Further Reading

What we have discussed here is just touching upon the topic. Hierarchical models in probabilistic modeling seem to be important research direction and modeling paradigm. Moreover, the technical details are also crucial for achieving state-of-the-art performance. I strongly suggest reading about NVAE [45], ResNet VAE [18], Ladder VAE [71], BIVA [44], and very deep VAEs [46] and compare various tricks and parameterizations used therein. These models share the same idea, but implementations vary significantly.

The research on hierarchical generative modeling is very up-to-date and develops very quickly. As a result, this is nearly impossible to mention even a fraction of interesting papers. I will mention only a few worth noticing papers:

- Pervez and Gavves [72] provides an insightful analysis about a potential problem with hierarchical VAEs, namely, the KL divergence term is closely related to the harmonics of the parameterizing function. In other words, using DNNs result in high-frequency components of the KL term and, eventually, to the posterior collapse. The authors propose to smooth the VAE by applying Ornstein–Uhlenbeck (OU) Semigroup. I refer to the original paper for details.
- Wu et al. [73] proposes greedy layer-wise learning of a hierarchical VAE. The authors used this idea in the context of video prediction. The main motivation for utilizing greedy layer-wise learning is a limited amount of computational resources. However, the idea of greedy layer-wise training has been extensively utilized in the past [66–68].
- Gatopoulos and Tomczak [25] discusses incorporating pre-defined transformations like downscaling into the model. The idea is to learn a reversed transformation to, e.g., downscaling in a stochastic manner. The resulting VAE has a set of auxiliary variables (e.g., downscaled versions of observables) and a set of latent variables that encode missing information in the auxiliary variables.

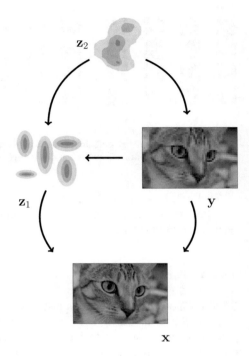

$$\mathbf{z}_2$$

$$\mathbf{z}_1 \qquad\qquad\qquad\qquad \mathbf{y}$$

$$\mathbf{x}$$

The hypothesis in such an approach is that learning a distribution over smaller or
already processed observable variables is easier and, thus, we can decompose the
problem into multiple problems of learning simpler distributions. A diagram for
this approach is presented in Fig. 4.22.

The beauty of the latent variable modeling paradigm is that we can play
with stochastic relationships among objects and, eventually, formulate a *useful*
representation of data. As we will see in the next blog posts, there are other
interesting classes of models that take advantage of diffusion models and energy
functions.

4.5.3   Diffusion-Based Deep Generative Models

4.5.3.1   Introduction

In Sect. 4.5, we discussed the issue of learning *useful* representations in latent
variable models, taking a closer look at hierarchical Variational Auto-Encoders. We
hypothesize that we can obtain *useful* data representations by applying a hierarchical
latent variable model. Moreover, highlighted a real problem in hierarchical VAEs of
the variational posterior collapsing to the prior, resulting in learning meaningless

Fig. 4.23 An example of applying a Gaussian diffusion to an image of a cat, x

representation. In other words, it seems that architecture with bottom-up variational posteriors (i.e., stochastic dependencies going from observables to the last latents) and top-down generative distributions seems to be a mediocre inductive bias and is rather troublesome to train. A potential solution is top-down VAEs. However, is there nothing we can do about the *vanilla* structure? As you may imagine, nothing is lost and some approaches take advantage of the bottom-up and the top-down structures. Here, we will look into the *diffusion-based deep generative models* (DDGMs) (a.k.a. *deep diffusion probabilistic models*) [74, 75].

DDGM could be briefly explained as hierarchical VAEs with the bottom-up path (i.e., the variational posteriors) defined by a diffusion process (e.g., a Gaussian diffusion) and the top-down path parameterized by DNNs (a reversed diffusion). Interestingly, the bottom-up path could be **fixed**, namely, it necessarily does not have any learnable parameters. An example of applying a Gaussian diffusion is presented in Fig. 4.23. Since the variational posteriors are fixed, we can think of them as adding Gaussian noise at each layer. Then, the final layer resembles Gaussian noise (see z_5 in Fig. 4.23). If we recall the discussion about a potential issue of the posterior collapse in hierarchical VAEs, this should not be a problem anymore. Why? Because we should get a standard Gaussian distribution in the last layer **by design**. Pretty neat, isn't it?

DDGMs have become extremely popular these days. They are appealing for at least two reasons:

1. They give amazing results for image synthesis [74, 76, 77], audio synthesis [78], and promising results for text synthesis [79, 80] while being relatively simple to implement.
2. They are closely related to stochastic differential equations and, thus, their theoretical properties seem to be especially of great interest [81–83].

There are two potential drawbacks though, namely:

1. DDGMs are unable (for now at least) to learn a representation.
2. Similarly to flow-based models, the dimensionality of input is kept across the whole model (i.e., there is no bottleneck on the way).

4.5.3.2 Model Formulation

Originally, deep diffusion probabilistic models were proposed in [75] and they took inspiration from non-equilibrium statistical physics. The main idea is to iteratively destroy the structure in data through a forward diffusion process and, afterward, to learn a reverse diffusion process to restore the structure in data. In a follow-up paper [74] recent developments in deep learning were used to train a powerful and flexible diffusion-based deep generative model that achieved SOTA results in the task of image synthesis. Here, we will abuse the original notation to make a clear connection between hierarchical latent variable models and DDGMs. As previously, we are interested in finding a distribution over data, $p_\theta(\mathbf{x})$; however, we assume an additional set of latent variables $\mathbf{z}_{1:T} = [\mathbf{z}_1, \dots, \mathbf{z}_T]$. The marginal likelihood is defined by integrating out all latents:

$$p_\theta(\mathbf{x}) = \int p_\theta(\mathbf{x}, \mathbf{z}_{1:T}) \, \mathrm{d}\mathbf{z}_{1:T}. \tag{4.86}$$

The joint distribution is modeled as a first-order Markov chain with Gaussian transitions, namely:

$$p_\theta(\mathbf{x}, \mathbf{z}_{1:T}) = p_\theta(\mathbf{x}|\mathbf{z}_1) \left(\prod_{i=1}^{T-1} p_\theta(\mathbf{z}_i|\mathbf{z}_{i+1}) \right) p_\theta(\mathbf{z}_T), \tag{4.87}$$

where $\mathbf{x} \in \mathbb{R}^D$ and $\mathbf{z}_i \in \mathbb{R}^D$ for $i = 1, \dots, T$. Please note that the latents have the same dimensionality as the observables. This is the same situation as in the case of flow-based models. We parameterize all distributions using DNNs.

So far, we have not introduced anything new! This is again a hierarchical latent variable model. As in the case of hierarchical VAEs, we can introduce a family of variational posteriors as follows:

$$Q_\phi(\mathbf{z}_{1:T}|\mathbf{x}) = q_\phi(\mathbf{z}_1|\mathbf{x}) \left(\prod_{i=2}^{T} q_\phi(\mathbf{z}_i|\mathbf{z}_{i-1}) \right). \tag{4.88}$$

The key point is how we define these distributions. Before, we used normal distributions parameterized by DNNs, but now we formulate them as the following Gaussian diffusion process [75]:

$$q_\phi(\mathbf{z}_i|\mathbf{z}_{i-1}) = \mathcal{N}(\mathbf{z}_i|\sqrt{1 - \beta_i}\mathbf{z}_{i-1}, \beta_i \mathbf{I}), \tag{4.89}$$

where $\mathbf{z}_0 = \mathbf{x}$. Notice that a single step of the diffusion, $q_\phi(\mathbf{z}_i|\mathbf{z}_{i-1})$, works in a relatively easy way. Namely, it takes the previously generated object $\mathbf{z}_{i-1}$, scales it by $\sqrt{1 - \beta_i}$, and then adds noise with variance β_i. To be even more explicit, we can write it using the reparameterization trick:

$$\mathbf{z}_i = \sqrt{1 - \beta_i}\mathbf{z}_{i-1} + \sqrt{\beta_i} \odot \epsilon, \tag{4.90}$$

where $\epsilon \sim \mathcal{N}(0, \mathbf{I})$. In principle, β_i could be learned by backpropagation; however, as noted by Sohl-Dickstein et al. [75] and Ho et al.[74], it could be fixed. For instance, [74] suggests to change it linearly from $\beta_1 = 10^{-4}$ to $\beta_T = 0.02$.

Since we realized that the difference between a DDGM and a hierarchical VAE lies in the definition of the variational posteriors and the dimensionality of the latents, but the whole construction is basically the same, we can predict what is the learning objective. Do you remember? Yes, it is ELBO! We can derive the ELBO as follows:

$$\ln p_\theta(\mathbf{x}) = \ln \int Q_\phi(\mathbf{z}_{1:T}|\mathbf{x}) \frac{p_\theta(\mathbf{x}, \mathbf{z}_{1:T})}{Q_\phi(\mathbf{z}_{1:T}|\mathbf{x})} \, d\mathbf{z}_{1:T}$$

$$\geq \mathbb{E}_{Q_\phi(\mathbf{z}_{1:T}|\mathbf{x})} \left[\ln p_\theta(\mathbf{x}|\mathbf{z}_1) + \sum_{i=1}^{T-1} \ln p_\theta(\mathbf{z}_i|\mathbf{z}_{i+1}) + \ln p_\theta(\mathbf{z}_T) + \right.$$

$$\left. - \sum_{i=2}^{T} \ln q_\phi(\mathbf{z}_i|\mathbf{z}_{i-1}) - \ln q_\phi(\mathbf{z}_1|\mathbf{x}) \right]$$

$$= \mathbb{E}_{Q_\phi(\mathbf{z}_{1:T}|\mathbf{x})} \left[\ln p_\theta(\mathbf{x}|\mathbf{z}_1) + \ln p_\theta(\mathbf{z}_1|\mathbf{z}_2) + \sum_{i=2}^{T-1} \ln p_\theta(\mathbf{z}_i|\mathbf{z}_{i+1}) + \ln p_\theta(\mathbf{z}_T) + \right.$$

$$\left. - \sum_{i=2}^{T-1} \ln q_\phi(\mathbf{z}_i|\mathbf{z}_{i-1}) - \ln q_\phi(\mathbf{z}_T|\mathbf{z}_{T-1}) - \ln q_\phi(\mathbf{z}_1|\mathbf{x}) \right]$$

$$= \mathbb{E}_{Q_\phi(\mathbf{z}_{1:T}|\mathbf{x})} \left[\ln p_\theta(\mathbf{x}|\mathbf{z}_1) + \sum_{i=2}^{T-1} \left(\ln p_\theta(\mathbf{z}_i|\mathbf{z}_{i+1}) - \ln q_\phi(\mathbf{z}_i|\mathbf{z}_{i-1}) \right) + \right.$$

$$+ \ln p_\theta(\mathbf{z}_T) - \ln q_\phi(\mathbf{z}_T|\mathbf{z}_{T-1}) +$$

$$\left. + \ln p_\theta(\mathbf{z}_1|\mathbf{z}_2) - \ln q_\phi(\mathbf{z}_1|\mathbf{x}) \right] \qquad (4.91)$$

$$\stackrel{df}{=} \mathcal{L}(\mathbf{x}; \theta, \phi).$$

We can rewrite the ELBO in terms of Kullback–Leibler divergences (note that we use the expected value with respect to $Q_\phi(\mathbf{z}_{-i}|\mathbf{x})$ to highlight that a proper variational posterior is used for the definition of the Kullback–Leibler divergence):

$$\mathcal{L}(\mathbf{x}; \theta, \phi) = \mathbb{E}_{Q_\phi(\mathbf{z}_{1:T}|\mathbf{x})} [\ln p_\theta(\mathbf{x}|\mathbf{z}_1)] +$$

$$- \sum_{i=2}^{T-1} \mathbb{E}_{Q_\phi(\mathbf{z}_{-i}|\mathbf{x})} \left[KL \left[q_\phi(\mathbf{z}_i|\mathbf{z}_{i-1}) || p_\theta(\mathbf{z}_i|\mathbf{z}_{i+1}) \right] \right] +$$

$$- \mathbb{E}_{Q_\phi(\mathbf{z}_{-T}|\mathbf{x})} \left[KL \left[q_\phi(\mathbf{z}_T|\mathbf{z}_{T-1}) || p_\theta(\mathbf{z}_T) \right] \right] +$$

$$- \mathbb{E}_{Q_\phi(\mathbf{z}_{-1}|\mathbf{x})} \left[KL \left[q_\phi(\mathbf{z}_1|\mathbf{x}) || p_\theta(\mathbf{z}_1|\mathbf{z}_2) \right] \right]. \tag{4.92}$$

Example 4.1 Let us take $T = 5$. This is not much (e.g., [74] uses $T = 1000$), but it is easier to explain the idea with a very specific model. Moreover, let us use a fixed $\beta_t \equiv \beta$. Then we have the following DDGM:

$$p_\theta(\mathbf{x}, \mathbf{z}_{1:5}) = p_\theta(\mathbf{x}|\mathbf{z}_1) p_\theta(\mathbf{z}_1|\mathbf{z}_2) p_\theta(\mathbf{z}_2|\mathbf{z}_3) p_\theta(\mathbf{z}_3|\mathbf{z}_4) p_\theta(\mathbf{z}_4|\mathbf{z}_5) p_\theta(\mathbf{z}_5), \tag{4.93}$$

and the variational posteriors:

$$Q_\phi(\mathbf{z}_{1:5}|\mathbf{x}) = q_\phi(\mathbf{z}_1|\mathbf{x}) q_\phi(\mathbf{z}_2|\mathbf{z}_1) q_\phi(\mathbf{z}_3|\mathbf{z}_2) q_\phi(\mathbf{z}_4|\mathbf{z}_3) q_\phi(\mathbf{z}_5|\mathbf{z}_4). \tag{4.94}$$

In the considered case, the ELBO takes the following form:

$$\begin{aligned}
\mathcal{L}(\mathbf{x}; \theta, \phi) = & \mathbb{E}_{Q_\phi(\mathbf{z}_{1:5}|\mathbf{x})} \left[\ln p_\theta(\mathbf{x}|\mathbf{z}_1) \right] + \\
& - \sum_{i=2}^{4} \mathbb{E}_{Q_\phi(\mathbf{z}_{-i}|\mathbf{x})} \left[KL \left[q_\phi(\mathbf{z}_i|\mathbf{z}_{i-1}) || p_\theta(\mathbf{z}_i|\mathbf{z}_{i+1}) \right] \right] + \\
& - \mathbb{E}_{Q_\phi(\mathbf{z}_{-i}|\mathbf{x})} \left[KL \left[q_\phi(\mathbf{z}_5|\mathbf{z}_4) || p_\theta(\mathbf{z}_5) \right] \right] + \\
& - \mathbb{E}_{Q_\phi(\mathbf{z}_{-i}|\mathbf{x})} \left[KL \left[q_\phi(\mathbf{z}_1|\mathbf{x}) || p_\theta(\mathbf{z}_1|\mathbf{z}_2) \right] \right],
\end{aligned} \tag{4.95}$$

where

$$p_\theta(\mathbf{z}_5) = \mathcal{N}(\mathbf{z}_5|0, \mathbf{I}). \tag{4.96}$$

The last interesting question is how to model inputs and, eventually, what distribution we should use to model $p(\mathbf{x}|\mathbf{z}_1)$. So far, we used the categorical distribution because pixels were integer-valued. However, for the DDGM, we assume that they are continuous and we will use a simple trick. We normalize our inputs to values between -1 and 1 and apply the Gaussian distribution with the unit variance and the mean being constrained to $[-1, 1]$ using the tanh non-linearity:

$$p(\mathbf{x}|\mathbf{z}_1) = \mathcal{N}(\mathbf{x}|\tanh(NN(\mathbf{z}_1)), \mathbf{I}), \tag{4.97}$$

where $NN(\mathbf{z}_1)$ is a neural network. As a result, since the variance is one, $\ln p(\mathbf{x}|\mathbf{z}_1) = -MSE(\mathbf{x}, \tanh(NN(\mathbf{z}_1))) + const$, so it is equivalent to the (negative) *Mean Squared Error*! I know, it is not a perfect way to do, but it is simple and it works. ∎

That's it! As you can see, there is no special magic here and we are ready to implement our DDGM. In fact, we can use the code of a hierarchical VAE and modify it accordingly. What is convenient about the DDGM is that the forward diffusion (i.e., the variational posteriors) are fixed and we need to sample from them, and only the reverse diffusion requires applying DDNs. But without any further mumbling, let us dive into the code!

4.5.3.3   Code

At this point, you might think that it is pretty complicated and a lot of math is involved here. However, if you followed our previous discussions on VAEs, it should be rather clear what we need to do here.

```
class DDGM(nn.Module):
    def __init__(self, p_dnns, decoder_net, beta, T, D):
        super(DDGM, self).__init__()

        print('DDGM by JT.')

        self.p_dnns = p_dnns # a list of sequentials; a single
        Sequential defines a DNN to parameterize a distribution p(z_i
        | z_i+1)

        self.decoder_net = decoder_net # the last DNN for p(x|z1)

        # other params
        self.D = D # the dimensionality of the inputs (necessary
        for sampling!)

        self.T = T # the number of steps

        self.beta = torch.FloatTensor([beta]) # the fixed
        variance of diffusion

        # The reparameterization trick for the Gaussian distribution
        @staticmethod
        def reparameterization(mu, log_var):
            std = torch.exp(0.5*log_var)
            eps = torch.randn_like(std)
            return mu + std * eps

        # The reparameterization trick for the Gaussian forward
        diffusion
        def reparameterization_gaussian_diffusion(self, x, i):
            return torch.sqrt(1. - self.beta) * x + torch.sqrt(self.
            beta) * torch.randn_like(x)

        def forward(self, x, reduction='avg'):
            # =====
            # Forward Diffusion
            # Please note that we just ''wander'' around in the space
            using Gaussian random walk.
            # We save all z's in a list
            zs = [self.reparameterization_gaussian_diffusion(x, 0)]
```

```
36    for i in range(1, self.T):
37        zs.append(self.reparameterization_gaussian_diffusion(
      zs[-1], i))
38
39    # =====
40    # Backward Diffusion
41    # We start with the last z and proceed to x.
42    # At each step, we calculate means and variances.
43    mus = []
44    log_vars = []
45
46    for i in range(len(self.p_dnns) - 1, -1, -1):
47        h = self.p_dnns[i](zs[i+1])
48        mu_i, log_var_i = torch.chunk(h, 2, dim=1)
49        mus.append(mu_i)
50        log_vars.append(log_var_i)
51
52    # The last step: outputting the means for x.
53    # NOTE: We assume the last distribution is Normal(x |
      tanh(NN(z_1)), 1)!
54    mu_x = self.decoder_net(zs[0])
55
56    # =====ELBO
57    # RE
58    # This is equivalent to - MSE(x, mu_x) + const
59    RE = log_standard_normal(x - mu_x).sum(-1)
60
61    # KL: We need to go through all the levels of latents
62    KL = (log_normal_diag(zs[-1], torch.sqrt(1. - self.beta)
      * zs[-1], torch.log(self.beta)) - log_standard_normal(zs[-1])
      ).sum(-1)
63
64    for i in range(len(mus)):
65        KL_i = (log_normal_diag(zs[i], torch.sqrt(1. - self.
      beta) * zs[i], torch.log(self.beta)) - log_normal_diag(zs[i],
      mus[i], log_vars[i])).sum(-1)
66
67        KL = KL + KL_i
68
69    # Final ELBO
70    if reduction == 'sum':
71        loss = -(RE - KL).sum()
72    else:
73        loss = -(RE - KL).mean()
74
75    return loss
76
77  # Sampling is the reverse diffusion with sampling at each
    step.
78  def sample(self, batch_size=64):
79      z = torch.randn([batch_size, self.D])
80      for i in range(len(self.p_dnns) - 1, -1, -1):
81          h = self.p_dnns[i](z)
82          mu_i, log_var_i = torch.chunk(h, 2, dim=1)
```

```
83          z = self.reparameterization(torch.tanh(mu_i),
       log_var_i)
84
85      mu_x = self.decoder_net(z)
86
87      return mu_x
88
89   # For sanity check, we also can sample from the forward
     diffusion.
90   # The result should resemble a white noise.
91   def sample_diffusion(self, x):
92      zs = [self.reparameterization_gaussian_diffusion(x, 0)]
93
94      for i in range(1, self.T):
95          zs.append(self.reparameterization_gaussian_diffusion(
       zs[-1], i))
96
97      return zs[-1]
```

Listing 4.13 A DDGM class

That's it! Now we are ready to run the full code (take a look at: https://github.com/jmtomczak/intro_dgm). After training our DDGM, we should obtain results like in Fig. 4.24.

4.5.3.4 Discussion

Extensions

Currently, DDGMs are very popular deep generative models. What we present here is very close to the original formulation of the DDGMs [75]. However, [74] introduced many interesting insights and improvements on the original idea, such as:

- Since the forward diffusion consists of Gaussian distributions and linear transformations of means, it is possible to analytically marginalize out intermediate steps, which yields:

$$q(\mathbf{z}_t | \mathbf{x}) = \mathcal{N}(\mathbf{z}_t | \sqrt{\bar{\alpha}_t} \mathbf{x}, (1 - \bar{\alpha}_t \mathbf{I}), \qquad (4.98)$$

where $\alpha_t = 1 - \beta_t$ and $\bar{\alpha}_t = \prod_{s=1}^{t} \alpha_t$. This is an extremely interesting result because we can sample $\mathbf{z}_t$ without sampling all intermediate steps!

- As a follow-up, we can calculate also the following distribution:

$$q(\mathbf{z}_{t-1} | \mathbf{z}_t, \mathbf{x}) = \mathcal{N}\left(\mathbf{z}_{t-1} | \tilde{\mu}_t(\mathbf{z}_t, \mathbf{x}), \tilde{\beta}_t \mathbf{I}\right), \qquad (4.99)$$

where:

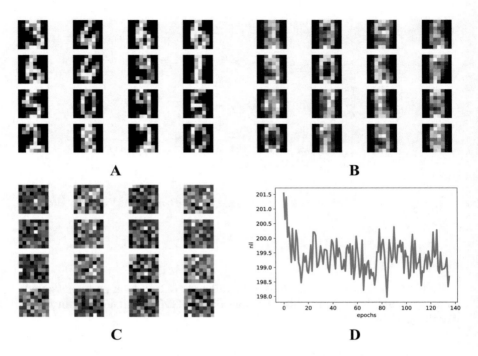

Fig. 4.24 An example of outcomes after the training: (**a**) Randomly selected real images. (**b**) Unconditional generations from the DDGM. (**c**) A visualization of the last stochastic level after applying the forward diffusion. As expected, the resulting images resemble pure noise. (**d**) An example of a validation curve for the ELBO

$$\tilde{\mu}_t\left(\mathbf{z}_t, \mathbf{x}\right) = \frac{\sqrt{\bar{\alpha}_{t-1}}\beta_t}{1 - \bar{\alpha}_t}\mathbf{x} + \frac{\sqrt{\alpha_t}\left(1 - \bar{\alpha}_{t-1}\right)}{1 - \bar{\alpha}_t}\mathbf{z}_t \tag{4.100}$$

and

$$\tilde{\beta}_t = \frac{1 - \bar{\alpha}_{t-1}}{1 - \bar{\alpha}_t}\beta_t. \tag{4.101}$$

Then, we can rewrite the ELBO as follows:

$$\mathcal{L}(\mathbf{x}; \theta, \phi) = \mathbb{E}_Q\Bigg[\underbrace{KL\left[q(\mathbf{z}_T|\mathbf{x})\| p(\mathbf{z}_T)\right]}_{L_T} +$$

$$+ \sum_{t>1} \underbrace{KL\left[q(\mathbf{z}_{t-1}|\mathbf{z}_t, \mathbf{x})\| p_\theta\left(\mathbf{z}_{t-1}|\mathbf{z}_t\right)\right]}_{L_{t-1}} +$$

$$\underbrace{-\log p_\theta\left(\mathbf{x} \mid \mathbf{z}_1\right)}_{L_0} \Bigg]. \tag{4.102}$$

Now, instead of differentiating all components of the objective, we can randomly pick L_t and treat it as the objective. Such an approach has a clear advantage: It does not require keeping all gradients in the memory! Instead, we update only one layer at a time. Since our training is stochastic anyway (remember that we typically use stochastic gradient descent), we can introduce this extra stochasticity during training. And the benefit is enormous because we can train extremely deep models, even with 1000 layers as in [74].

- If you play a little with the code here, you may notice that training the reverse diffusion is pretty problematic. Why? Because by adding extra layers of latents, we add additional KL-terms. In the case of far from perfect models, each KL-term will be strictly greater than 0 and, thus, we will increase the ELBO with each additional step of stochasticity. Therefore, it is so important to be smart about formulating reverse diffusion. [74] again provides very interesting insight! We skip here the full reasoning, but it turns out that to make the model $p_\theta (\mathbf{z}_{t-1} \mid \mathbf{z}_t) = \mathcal{N} (\mathbf{z}_{t-1} | \mu_\theta (\mathbf{z}_t), \sigma_t^2 \mathbf{I})$ more powerful, $\mu_\theta (\mathbf{z}_t)$ should be as close as possible to $\tilde{\mu}_t (\mathbf{z}_t, \mathbf{x})$. Following the derivation in [74], we get

$$\mu_\theta (\mathbf{z}_t) = \frac{1}{\sqrt{\alpha_t}} \left(\mathbf{z}_t - \frac{\beta_t}{\sqrt{1 - \bar{\alpha}_t}} \epsilon_\theta (\mathbf{z}_t) \right),$$

where $\epsilon_\theta (\mathbf{z}_t)$ is parameterized by a DNN and it aims for estimating the noise from $\mathbf{z}_t$.

- Even further, each L_t could be simplified to:

$$L_{t,\text{simple}} = \mathbb{E}_{t,\mathbf{x}_0,\epsilon} \left[\left\| \epsilon - \epsilon_\theta \left(\sqrt{\bar{\alpha}_t} \mathbf{x}_0 + \sqrt{1 - \bar{\alpha}_t} \epsilon, t \right) \right\|^2 \right].$$

Ho et al. [74] provides empirical results that such an objective could be beneficial for training and the final quality of synthesized images.

There were many follow-ups on [74], we mention only a few of them here:

- *Improving DDGMs*: Nichol and Dhariwal [84] introduces further tricks on improving training stability and performance of DDGMs by learning the covariance matrices in the reverse diffusion, proposing a different noise schedule, among others. Interestingly, the authors of [76] propose to extend the observables (i.e., pixels) and latents with Fourier features as additional channels. The rationale behind this is that the high frequency features allow neural networks to cope better with noise. Moreover, they introduce a new noise scheduling and an application of certain functions to improve the numerical stability of the forward diffusion process.
- *Sampling speed-up*: Kong and Ping [85] and Watson et al. [86] focus on speeding up the sampling process.
- *Superresolution*: Saharia et al. [77] uses DDGMs for the task of superresolution.

- *Connection to score-based generative models*: It turns out that score-based generative models [87] are closely related to DDGMs as indicated by Song and Kingma [88] and Song and Ermon [87]. This perspective gives a neat connection between DDGMs and stochastic differential equations [81, 83].
- *Variational perspective on DDGMs*: There is a nice variational perspective on DDGMs [76, 81] that gives an intuitive interpretation of DDGMs and allows achieving astonishing results on the image synthesis task. It is worth to study [76] further because there are a lot of interesting improvements presented therein.
- *Discrete DDGMs*: So far, DDGMs are mainly focused on continuous spaces. [79, 80] propose DDGMs on discrete spaces.
- *DDGMs for audio*: Kong et al. [78] proposes to use DDGMs for audio synthesis.
- *DDGMs as priors in VAEs*: Vahdat et al. [89] and Wehenkel and Louppe [90] propose to use DDGMs as flexible priors in VAEs.

DDGMs vs. VAEs vs. Flows

In the end, it is worth making a comparison of DDGMs with VAEs and flow-based models. In Table 4.1, we provide a comparison based on rather arbitrary criteria:

- Whether the training procedure is stable or not
- Whether the likelihood could be calculated
- Whether a reconstruction is possible
- Whether a model is invertible
- Whether the latent representation could be lower dimensional than the input space (i.e., a bottleneck in a model)

The three models share a lot of similarities. Overall, training is rather stable even though numerical issues could arise in all models. Hierarchical VAEs could be seen as a generalization of DDGMs. There is an open question of whether it is indeed more beneficial to use fixed variational posteriors by sacrificing the possibility of having a bottleneck. There is also a connection between flows and DDGMs. Both classes of models aim for going from data to noise. Flows do that by applying invertible transformations, while DDGMs accomplish that by a diffusion process. In flows, we know the inverse but we pay the price of calculating the Jacobian-determinant, while DDGMs require flexible parameterizations of the reverse diffusion but there are no extra strings attached. Looking into connections among these models is definitely an interesting research line.

Table 4.1 A comparison among DDGMs, VAEs, and Flows

Model	Training	Likelihood	Reconstruction	Invertible	Bottleneck (latents)
DDGMs	Stable	Approximate	Difficult	No	No
VAEs	Stable	Approximate	Easy	No	Possible
Flows	Stable	Exact	Easy	Yes	No

References

1. Christopher M Bishop. *Pattern recognition and machine learning.* Springer, 2006.
2. Michael E Tipping and Christopher M Bishop. Probabilistic principal component analysis. *Journal of the Royal Statistical Society: Series B (Statistical Methodology)*, 61(3):611–622, 1999.
3. Christophe Andrieu, Nando De Freitas, Arnaud Doucet, and Michael I Jordan. An introduction to MCMC for machine learning. *Machine learning*, 50(1):5–43, 2003.
4. Michael I Jordan, Zoubin Ghahramani, Tommi S Jaakkola, and Lawrence K Saul. An introduction to variational methods for graphical models. *Machine learning*, 37(2):183–233, 1999.
5. Yoon Kim, Sam Wiseman, Andrew Miller, David Sontag, and Alexander Rush. Semi-amortized variational autoencoders. In *International Conference on Machine Learning*, pages 2678–2687. PMLR, 2018.
6. Diederik P Kingma and Max Welling. Auto-encoding variational Bayes. *arXiv preprint arXiv:1312.6114*, 2013.
7. Danilo Jimenez Rezende, Shakir Mohamed, and Daan Wierstra. Stochastic backpropagation and approximate inference in deep generative models. In *International conference on machine learning*, pages 1278–1286. PMLR, 2014.
8. Luc Devroye. Random variate generation in one line of code. In *Proceedings Winter Simulation Conference*, pages 265–272. IEEE, 1996.
9. Diederik Kingma and Max Welling. Efficient gradient-based inference through transformations between Bayes nets and neural nets. In *International Conference on Machine Learning*, pages 1782–1790. PMLR, 2014.
10. Alexander Alemi, Ben Poole, Ian Fischer, Joshua Dillon, Rif A Saurous, and Kevin Murphy. Fixing a broken ELBO. In *International Conference on Machine Learning*, pages 159–168. PMLR, 2018.
11. Samuel Bowman, Luke Vilnis, Oriol Vinyals, Andrew Dai, Rafal Jozefowicz, and Samy Bengio. Generating sentences from a continuous space. In *Proceedings of The 20th SIGNLL Conference on Computational Natural Language Learning*, pages 10–21, 2016.
12. Danilo Jimenez Rezende and Fabio Viola. Taming VAEs. *arXiv preprint arXiv:1810.00597*, 2018.
13. Eric Nalisnick, Akihiro Matsukawa, Yee Whye Teh, Dilan Gorur, and Balaji Lakshmi-narayanan. Do deep generative models know what they don't know? In *International Conference on Learning Representations*, 2018.
14. Charline Le Lan and Laurent Dinh. Perfect density models cannot guarantee anomaly detection. *arXiv preprint arXiv:2012.03808*, 2020.
15. Yuri Burda, Roger Grosse, and Ruslan Salakhutdinov. Importance weighted autoencoders. *arXiv preprint arXiv:1509.00519*, 2015.
16. Rianne Van Den Berg, Leonard Hasenclever, Jakub M Tomczak, and Max Welling. Sylvester normalizing flows for variational inference. In *34th Conference on Uncertainty in Artificial Intelligence 2018, UAI 2018*, pages 393–402. Association For Uncertainty in Artificial Intelligence (AUAI), 2018.
17. Emiel Hoogeboom, Victor Garcia Satorras, Jakub M Tomczak, and Max Welling. The convolution exponential and generalized Sylvester flows. *arXiv preprint arXiv:2006.01910*, 2020.
18. Durk P Kingma, Tim Salimans, Rafal Jozefowicz, Xi Chen, Ilya Sutskever, and Max Welling. Improved variational inference with inverse autoregressive flow. *Advances in Neural Information Processing Systems*, 29:4743–4751, 2016.
19. Danilo Rezende and Shakir Mohamed. Variational inference with normalizing flows. In *International Conference on Machine Learning*, pages 1530–1538. PMLR, 2015.
20. Jakub M Tomczak and Max Welling. Improving variational auto-encoders using householder flow. *arXiv preprint arXiv:1611.09630*, 2016.

21. Jakub M Tomczak and Max Welling. Improving variational auto-encoders using convex combination linear inverse autoregressive flow. *arXiv preprint arXiv:1706.02326*, 2017.
22. Ishaan Gulrajani, Kundan Kumar, Faruk Ahmed, Adrien Ali Taiga, Francesco Visin, David Vazquez, and Aaron Courville. PixelVAE: A latent variable model for natural images. *arXiv preprint arXiv:1611.05013*, 2016.
23. Jakub Tomczak and Max Welling. VAE with a VampPrior. In *International Conference on Artificial Intelligence and Statistics*, pages 1214–1223. PMLR, 2018.
24. Xi Chen, Diederik P Kingma, Tim Salimans, Yan Duan, Prafulla Dhariwal, John Schulman, Ilya Sutskever, and Pieter Abbeel. Variational lossy autoencoder. *arXiv preprint arXiv:1611.02731*, 2016.
25. Ioannis Gatopoulos and Jakub M Tomczak. Self-supervised variational auto-encoders. *Entropy*, 23(6):747, 2021.
26. Amirhossein Habibian, Ties van Rozendaal, Jakub M Tomczak, and Taco S Cohen. Video compression with rate-distortion autoencoders. In *Proceedings of the IEEE/CVF International Conference on Computer Vision*, pages 7033–7042, 2019.
27. Matthias Bauer and Andriy Mnih. Resampled priors for variational autoencoders. In *The 22nd International Conference on Artificial Intelligence and Statistics*, pages 66–75. PMLR, 2019.
28. Diederik P Kingma, Danilo J Rezende, Shakir Mohamed, and Max Welling. Semi-supervised learning with deep generative models. In *Proceedings of the 27th International Conference on Neural Information Processing Systems*, pages 3581–3589, 2014.
29. Christos Louizos, Kevin Swersky, Yujia Li, Max Welling, and Richard Zemel. The variational fair autoencoder. *arXiv preprint arXiv:1511.00830*, 2015.
30. Maximilian Ilse, Jakub M Tomczak, Christos Louizos, and Max Welling. DIVA: Domain invariant variational autoencoders. In *Medical Imaging with Deep Learning*, pages 322–348. PMLR, 2020.
31. Charles Blundell, Julien Cornebise, Koray Kavukcuoglu, and Daan Wierstra. Weight uncertainty in neural network. In *International Conference on Machine Learning*, pages 1613–1622. PMLR, 2015.
32. Wengong Jin, Regina Barzilay, and Tommi Jaakkola. Junction tree variational autoencoder for molecular graph generation. In *International Conference on Machine Learning*, pages 2323–2332. PMLR, 2018.
33. Tim R Davidson, Luca Falorsi, Nicola De Cao, Thomas Kipf, and Jakub M Tomczak. Hyperspherical variational auto-encoders. In *34th Conference on Uncertainty in Artificial Intelligence 2018, UAI 2018*, pages 856–865. Association For Uncertainty in Artificial Intelligence (AUAI), 2018.
34. Tim R Davidson, Jakub M Tomczak, and Efstratios Gavves. Increasing expressivity of a hyperspherical VAE. *arXiv preprint arXiv:1910.02912*, 2019.
35. Emile Mathieu, Charline Le Lan, Chris J Maddison, Ryota Tomioka, and Yee Whye Teh. Continuous Hierarchical Representations with Poincaré Variational Auto-Encoders. *arXiv preprint arXiv:1901.06033*, 2019.
36. Eric Jang, Shixiang Gu, and Ben Poole. Categorical reparameterization with Gumbel-Softmax. *arXiv preprint arXiv:1611.01144*, 2016.
37. C Maddison, A Mnih, and Y Teh. The concrete distribution: A continuous relaxation of discrete random variables. In *Proceedings of the international conference on learning Representations*. International Conference on Learning Representations, 2017.
38. Emile van Krieken, Jakub M Tomczak, and Annette ten Teije. Storchastic: A framework for general stochastic automatic differentiation. *Advances in Neural Information Processing Systems*, 2021.
39. Junxian He, Daniel Spokoyny, Graham Neubig, and Taylor Berg-Kirkpatrick. Lagging inference networks and posterior collapse in variational autoencoders. *arXiv preprint arXiv:1901.05534*, 2019.
40. Adji B Dieng, Yoon Kim, Alexander M Rush, and David M Blei. Avoiding latent variable collapse with generative skip models. In *The 22nd International Conference on Artificial Intelligence and Statistics*, pages 2397–2405. PMLR, 2019.

41. Adji B Dieng, Dustin Tran, Rajesh Ranganath, John Paisley, and David M Blei. Variational Inference via χ-Upper Bound Minimization. In *Proceedings of the 31st International Conference on Neural Information Processing Systems*, pages 2729–2738, 2017.
42. Irina Higgins, Loic Matthey, Arka Pal, Christopher Burgess, Xavier Glorot, Matthew Botvinick, Shakir Mohamed, and Alexander Lerchner. beta-VAE: Learning basic visual concepts with a constrained variational framework. *ICLR*, 2016.
43. Partha Ghosh, Mehdi SM Sajjadi, Antonio Vergari, Michael Black, and Bernhard Scholkopf. From variational to deterministic autoencoders. In *International Conference on Learning Representations*, 2019.
44. Lars Maaløe, Marco Fraccaro, Valentin Liévin, and Ole Winther. BIVA: A Very Deep Hierarchy of Latent Variables for Generative Modeling. In *NeurIPS*, 2019.
45. Arash Vahdat and Jan Kautz. NVAE: A deep hierarchical variational autoencoder. *arXiv preprint arXiv:2007.03898*, 2020.
46. Rewon Child. Very deep VAEs generalize autoregressive models and can outperform them on images. *arXiv preprint arXiv:2011.10650*, 2020.
47. Alireza Makhzani, Jonathon Shlens, Navdeep Jaitly, Ian Goodfellow, and Brendan Frey. Adversarial autoencoders. *arXiv preprint arXiv:1511.05644*, 2015.
48. Matthew D Hoffman and Matthew J Johnson. ELBO surgery: Yet another way to carve up the variational evidence lower bound. In *Workshop in Advances in Approximate Bayesian Inference, NIPS*, volume 1, page 2, 2016.
49. Ricky TQ Chen, Xuechen Li, Roger Grosse, and David Duvenaud. Isolating sources of disentanglement in VAEs. In *Proceedings of the 32nd International Conference on Neural Information Processing Systems*, pages 2615–2625, 2018.
50. Frantzeska Lavda, Magda Gregorová, and Alexandros Kalousis. Data-dependent conditional priors for unsupervised learning of multimodal data. *Entropy*, 22(8):888, 2020.
51. Shuyu Lin and Ronald Clark. Ladder: Latent data distribution modelling with a generative prior. *arXiv preprint arXiv:2009.00088*, 2020.
52. Christopher M Bishop, Markus Svensén, and Christopher KI Williams. GTM: The generative topographic mapping. *Neural computation*, 10(1):215–234, 1998.
53. Christian H Bischof and Xiaobai Sun. On orthogonal block elimination. *Preprint MCS-P450-0794, Mathematics and Computer Science Division, Argonne National Laboratory*, page 4, 1994.
54. Xiaobai Sun and Christian Bischof. A basis-kernel representation of orthogonal matrices. *SIAM journal on matrix analysis and applications*, 16(4):1184–1196, 1995.
55. Alston S Householder. Unitary triangularization of a nonsymmetric matrix. *Journal of the ACM (JACM)*, 5(4):339–342, 1958.
56. Leonard Hasenclever, Jakub Tomczak, Rianne van den Berg, and Max Welling. Variational inference with orthogonal normalizing flows. 2017.
57. Åke Björck and Clazett Bowie. An iterative algorithm for computing the best estimate of an orthogonal matrix. *SIAM Journal on Numerical Analysis*, 8(2):358–364, 1971.
58. Zdislav Kovarik. Some iterative methods for improving orthonormality. *SIAM Journal on Numerical Analysis*, 7(3):386–389, 1970.
59. Gary Ulrich. Computer generation of distributions on the m-sphere. *Journal of the Royal Statistical Society: Series C (Applied Statistics)*, 33(2):158–163, 1984.
60. Christian Naesseth, Francisco Ruiz, Scott Linderman, and David Blei. Reparameterization gradients through acceptance-rejection sampling algorithms. In *Artificial Intelligence and Statistics*, pages 489–498. PMLR, 2017.
61. Yoshua Bengio, Aaron Courville, and Pascal Vincent. Representation learning: A review and new perspectives. *IEEE transactions on pattern analysis and machine intelligence*, 35(8):1798–1828, 2013.
62. Ferenc Huszár. Is maximum likelihood useful for representation learning?
63. Mary Phuong, Max Welling, Nate Kushman, Ryota Tomioka, and Sebastian Nowozin. The mutual autoencoder: Controlling information in latent code representations.

64. Samarth Sinha and Adji B Dieng. Consistency regularization for variational auto-encoders. *arXiv preprint arXiv:2105.14859*, 2021.
65. Jakub M Tomczak. Learning informative features from restricted Boltzmann machines. *Neural Processing Letters*, 44(3):735–750, 2016.
66. Yoshua Bengio. *Learning deep architectures for AI*. Now Publishers Inc, 2009.
67. Ruslan Salakhutdinov. Learning deep generative models. *Annual Review of Statistics and Its Application*, 2:361–385, 2015.
68. Ruslan Salakhutdinov and Geoffrey Hinton. Deep Boltzmann machines. In *Artificial intelligence and statistics*, pages 448–455. PMLR, 2009.
69. Andrew Gelman, John B Carlin, Hal S Stern, and Donald B Rubin. *Bayesian data analysis*. Chapman and Hall/CRC, 1995.
70. Lars Maaløe, Marco Fraccaro, and Ole Winther. Semi-supervised generation with cluster-aware generative models. *arXiv preprint arXiv:1704.00637*, 2017.
71. Casper Kaae Sønderby, Tapani Raiko, Lars Maaløe, Søren Kaae Sønderby, and Ole Winther. Ladder variational autoencoders. *Advances in Neural Information Processing Systems*, 29:3738–3746, 2016.
72. Adeel Pervez and Efstratios Gavves. Spectral smoothing unveils phase transitions in hierarchical variational autoencoders. In *International Conference on Machine Learning*, pages 8536–8545. PMLR, 2021.
73. Bohan Wu, Suraj Nair, Roberto Martin-Martin, Li Fei-Fei, and Chelsea Finn. Greedy hierarchical variational autoencoders for large-scale video prediction. In *Proceedings of the IEEE/CVF Conference on Computer Vision and Pattern Recognition*, pages 2318–2328, 2021.
74. Jonathan Ho, Ajay Jain, and Pieter Abbeel. Denoising diffusion probabilistic models. *arXiv preprint arXiv:2006.11239*, 2020.
75. Jascha Sohl-Dickstein, Eric Weiss, Niru Maheswaranathan, and Surya Ganguli. Deep unsupervised learning using nonequilibrium thermodynamics. In *International Conference on Machine Learning*, pages 2256–2265. PMLR, 2015.
76. Diederik P Kingma, Tim Salimans, Ben Poole, and Jonathan Ho. Variational diffusion models. *arXiv preprint arXiv:2107.00630*, 2021.
77. Chitwan Saharia, Jonathan Ho, William Chan, Tim Salimans, David J Fleet, and Mohammad Norouzi. Image super-resolution via iterative refinement. *arXiv preprint arXiv:2104.07636*, 2021.
78. Zhifeng Kong, Wei Ping, Jiaji Huang, Kexin Zhao, and Bryan Catanzaro. DiffWave: A versatile diffusion model for audio synthesis. In *International Conference on Learning Representations*, 2020.
79. Jacob Austin, Daniel Johnson, Jonathan Ho, Danny Tarlow, and Rianne van den Berg. Structured denoising diffusion models in discrete state-spaces. *arXiv preprint arXiv:2107.03006*, 2021.
80. Emiel Hoogeboom, Didrik Nielsen, Priyank Jaini, Patrick Forré, and Max Welling. Argmax flows and multinomial diffusion: Towards non-autoregressive language models. *arXiv preprint arXiv:2102.05379*, 2021.
81. Chin-Wei Huang, Jae Hyun Lim, and Aaron Courville. A variational perspective on diffusion-based generative models and score matching. *arXiv preprint arXiv:2106.02808*, 2021.
82. Yang Song, Jascha Sohl-Dickstein, Diederik P Kingma, Abhishek Kumar, Stefano Ermon, and Ben Poole. Score-based generative modeling through stochastic differential equations. In *International Conference on Learning Representations*, 2020.
83. Belinda Tzen and Maxim Raginsky. Neural stochastic differential equations: Deep latent Gaussian models in the diffusion limit. *arXiv preprint arXiv:1905.09883*, 2019.
84. Alex Nichol and Prafulla Dhariwal. Improved denoising diffusion probabilistic models. *arXiv preprint arXiv:2102.09672*, 2021.
85. Zhifeng Kong and Wei Ping. On fast sampling of diffusion probabilistic models. *arXiv preprint arXiv:2106.00132*, 2021.
86. Daniel Watson, Jonathan Ho, Mohammad Norouzi, and William Chan. Learning to efficiently sample from diffusion probabilistic models. *arXiv preprint arXiv:2106.03802*, 2021.

87. Yang Song and Stefano Ermon. Generative modeling by estimating gradients of the data distribution. *arXiv preprint arXiv:1907.05600*, 2019.
88. Yang Song and Diederik P Kingma. How to train your energy-based models. *arXiv preprint arXiv:2101.03288*, 2021.
89. Arash Vahdat, Karsten Kreis, and Jan Kautz. Score-based generative modeling in latent space. *arXiv preprint arXiv:2106.05931*, 2021.
90. Antoine Wehenkel and Gilles Louppe. Diffusion priors in variational autoencoders. In *ICML Workshop on Invertible Neural Networks, Normalizing Flows, and Explicit Likelihood Models*, 2021.

Chapter 5
Hybrid Modeling

5.1 Introduction

In Chap. 1, I tried to convince you that learning the conditional distribution $p(y|\mathbf{x})$ is not enough and, instead, we should focus on the joint distribution $p(\mathbf{x}, y)$ factorized as follows:

$$p(\mathbf{x}, y) = p(y|\mathbf{x})p(\mathbf{x}). \tag{5.1}$$

Why? Let me remind you my reasoning. The conditional $p(y|\mathbf{x})$ does not allow us to say anything about $\mathbf{x}$ but, instead, it will give its best to provide a decision. As a result, I can provide an object that has never been observed so far, and $p(y|\mathbf{x})$ could still be pretty certain about its decision (i.e., assigning high probability to one class). On the other hand, once we have trained $p(\mathbf{x})$, we should be able to, at least in theory, access the probability of the given object. And, eventually, determine whether our decision is reliable or not.

In the previous chapters, we completely focused on answering the question on how to learn $p(\mathbf{x})$ alone. Since we had in mind the necessity of using it for evaluating the probability, we discussed only the likelihood-based models, namely the autoregressive models (ARMs), the flow-based models (flows), and the Variational Auto-Encoders (VAEs). Now, the naturally arising question is how to use a deep generative model together with a classifier (or a regressor). Let us focus on a classification task for simplicity and think of possible approaches.

5.1.1 Approach 1: Let's Be Naive!

Let us start with some easy, naive almost approach. In the most straightforward way, we can train $p(y|\mathbf{x})$ and $p(\mathbf{x})$ separately. And that is it, we have a classifier,

J. M. Tomczak, *Deep Generative Modeling*,
https://doi.org/10.1007/978-3-030-93158-2_5

Fig. 5.1 A naive approach to
learning the joint distribution
by considering both
distributions separately

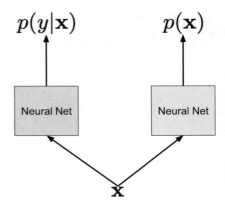

and a marginal distribution over objects. This approach is schematically presented
in Fig. 5.1 where we use different colors (purple and blue) to highlight that we use
two different neural networks to parameterize the two distributions.

Taking the logarithm of the joint distribution yields

$$\ln p(\mathbf{x}, y) = \ln p_\alpha(y|\mathbf{x}) + \ln p_\beta(\mathbf{x}), \tag{5.2}$$

where α and β denote parameterizations of both distributions (i.e., neural networks).
Once we start training and calculate gradients with respect to α and β, we clearly
see that we get

$$\nabla_\alpha \ln p(\mathbf{x}, y) = \nabla_\alpha \ln p_\alpha(y|\mathbf{x}) + \underbrace{\nabla_\alpha \ln p_\beta(\mathbf{x})}_{=0}, \tag{5.3}$$

because $\ln p_\beta(\mathbf{x})$ is not dependent on α, and

$$\nabla_\beta \ln p(\mathbf{x}, y) = \underbrace{\nabla_\beta \ln p_\alpha(y|\mathbf{x})}_{=0} + \nabla_\beta \ln p_\beta(\mathbf{x}), \tag{5.4}$$

because $\ln p_\alpha(y|\mathbf{x})$ does not depend on β.

In other words, we can simply first train $p_\alpha(y|\mathbf{x})$ using all data with labels, and
then train $p_\beta(\mathbf{x})$ using all available data. So what is a potential pitfall with this
approach? Intuitively, we can say that there is no guarantee that both distributions
treat $\mathbf{x}$ in the same manner and, thus, could introduce some errors. Moreover, due
to the stochasticity during training, there is no information flow between random
variables $\mathbf{x}$ and y and, as a result, the neural networks seek for own (local)
minima. To use a metaphor, they are like two wings of a bird that move in total
separation, completely asynchronously. Moreover, training both models separately
is also inefficient. We must use two different neural networks, with no weight

sharing. Since training is stochastic, we really could worry about potential bad local optima and our worries are even doubled now.

Would such an approach fail? Well, there is no simple answer to this question. Probably, it could work pretty well even, but it might lead to models far from optimal ones. Either way, who does like being unclear about training models? At least not me.

5.1.2 Approach 2: Shared Parameterization!

Alright, so since I whine about sharing the parameterization, it is obvious that the second approach uses (drums here)... a shared parameterization! To be more precise, a partially shared parameterization assumes that there is a neural network that processes $\mathbf{x}$ and then its output is fed to two neural networks: one for the classifier, and one for the marginal distribution over $\mathbf{x}$'s. An example of this approach is depicted in Fig. 5.2 (the shared neural network is depicted in purple).

Now, taking the logarithm of the joint distribution gives

$$\ln p(\mathbf{x}, y) = \ln p_{\alpha,\gamma}(y|\mathbf{x}) + \ln p_{\beta,\gamma}(\mathbf{x}), \tag{5.5}$$

where, it is worth highlighting, both distributions partially share the parameterization γ (i.e., the purple neural network in Fig. 5.2). As a result, during training, there is an obvious information sharing between $\mathbf{x}$ and y! Intuitively, both distributions operate on a processed $\mathbf{x}$ in the same manner, and then this representation is specialized to give probabilities for classes and objects.

Again, one might ask what is this all fuzz about?! Well, first of all, now the two distributions are tightly connected. Like in the metaphor of a bird used before, both wings can move together, in a synchronized fashion. Second, from the

Fig. 5.2 An approach to learning the joint distribution by using a partially shared parameterization

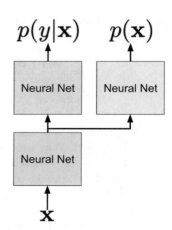

optimization perspective, the gradients flow through the γ network, and, thus, it contains information about both $\mathbf{x}$ and y. This may greatly help in finding a better solution.

5.2   Hybrid Modeling

At the first glance, there is nothing wrong in learning using the training objective expressed as

$$\ln p(\mathbf{x}, y) = \ln p_{\alpha,\gamma}(y|\mathbf{x}) + \ln p_{\beta,\gamma}(\mathbf{x}). \tag{5.6}$$

However, let us think about dimensionalities of y and $\mathbf{x}$. For instance, if y is binary, then we have one single bit representing a class label. For a binary vector of $\mathbf{x}$, we have D bits. Hence, there is a clear discrepancy in scales! Let us take a look at the gradient with respect to γ first, namely:

$$\nabla_\gamma \ln p(\mathbf{x}, y) = \nabla_\gamma \ln p_{\alpha,\gamma}(y|\mathbf{x}) + \nabla_\gamma \ln p_{\beta,\gamma}(\mathbf{x}). \tag{5.7}$$

If we think about it, during training, the γ network obtains a much stronger signal from $\ln p_{\beta,\gamma}(\mathbf{x})$. Following our example of binary variables, let us assume that our neural nets return all probabilities equal 0.5, so for the independent Bernoulli variables we get

$$\ln Bern(y|0.5) = y \ln 0.5 + (1 - y) \ln 0.5$$
$$= - \ln 2,$$

where we use the property of the logarithm ($\ln 0.5 = \ln 2^{-1} = - \ln 2$) and it does not matter what is the value of y because the neural network returns 0.5 for $y = 0$ and $y = 1$. Similarly, for $\mathbf{x}$ we get

$$\ln \prod_{d=1}^{D} Bern(x_d|0.5) = \sum_{d=1}^{D} \ln Bern(x_d|0.5)$$
$$= -D \ln 2.$$

Therefore, we see that the $\ln p_{\beta,\gamma}(\mathbf{x})$ part is D-times stronger than the $\ln p_{\alpha,\gamma}(y|\mathbf{x})$ part! How does it influence the final gradients during training? Try to visualize a bar of height $\ln 2$ and the other that is D-times higher. Now, imagine these bars "flow" through γ. Do you see it? Yes, the γ neural network will obtain more information from the marginal distribution and this information could cripple the classification part. In other words, our final model will be always **biased towards the marginal part**. Can we do something about it? Fortunately, yes!

Fig. 5.3 Hybrid modeling using invertible neural networks and flow-based models

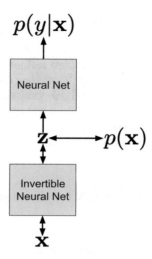

In [1] it was proposed to consider the convex combination of $\ln p(y|\mathbf{x})$ and $\ln p(\mathbf{x})$ as the objective function, namely:

$$\mathcal{L}(\mathbf{x}, y; \lambda) = (1 - \lambda) \ln p(y|\mathbf{x}) + \lambda \ln p(\mathbf{x}), \qquad (5.8)$$

where $\lambda \in [0, 1]$. Unfortunately, this weighting scheme is not derived from a well-defined distribution and it breaks the elegance of the likelihood-based approach. However, if you do not mind being inelegant, then this approach should work well!

A different approach is proposed in [2] where only $\ln p(\mathbf{x})$ is weighted:

$$\ell(\mathbf{x}, y; \lambda) = \ln p(y|\mathbf{x}) + \lambda \ln p(\mathbf{x}), \qquad (5.9)$$

where $\lambda \geq 0$. This kind of weighting was proposed in various forms before (e.g., see [3, 4]). Still, the fudge factor λ is not derived from a probabilistic perspective. However, [2] argues that we can interpret λ as a way of encouraging robustness to input variations. They also mention that scaling $\ln p(\mathbf{x})$ can be seen as a Jacobian-based regularization penalty. It is still not a valid distribution (because it is equivalent to $p(\mathbf{x})^{\lambda}$), but at least we can provide some interpretations.

In [2], the hybrid modeling idea has been pursued with $p(\mathbf{x})$ being modeled by flows (in the paper they used GLOW [5]) and then, the resulting latents $\mathbf{z}$ were used as the input to the classifier. In other words, a flow-based model is used for $p(\mathbf{x})$ and the invertible neural network (e.g., consisting of coupling layers) is shared with the classifier. Then, the final layers on top of the invertible neural network are used to make a decision y. The objective function is $\ell(\mathbf{x}, y; \lambda)$ as defined in Eq. (5.9). The approach is schematically presented in Fig. 5.3.

There are a couple of interesting properties of this approach. First, we can use the invertible neural network for both the generative and discriminative parts of the model. Hence, the flow-based model is well-informed about the label. Second, the weighting λ allows controlling whether the model is *more* discriminative or *more* generative. Third, we can use **any** flow-based model! GLOW was used in [2], however, [6] used residual flows, and [7] applied invertible DenseNets. Fourth, as presented by [2], we can use **any** classifier (or regressor), e.g., Bayesian classifiers.

A potential drawback of this approach lies in the necessity of determining λ. This is an extra hyperparameter that requires tuning. Moreover, as noticed in previous papers [2, 6, 7], the value of λ drastically changes the performance of the model from discriminative to generative. That is an open question how to deal with that.

5.3   Let's Implement It!

No it is time to be more specific and formulate a hybrid model. Let us start with the classifier and consider a fully-connected neural network to model the conditional distribution $p(y|\mathbf{x})$, namely:

$$\mathbf{z} \rightarrow \text{Linear}(D, M) \rightarrow \text{ReLU} \rightarrow \text{Linear}(M, M) \rightarrow \text{ReLU} \rightarrow$$
$$\rightarrow \text{Linear}(M, K) \rightarrow \text{Softmax},$$

where D is the dimensionality of $\mathbf{x}$ and K is the number of classes. The softmax gives us probabilities for each class. Remember that $\mathbf{z} = f^{-1}(\mathbf{x})$, where f is an invertible neural network.

In our example, we use the classifier, so we should take the categorical distribution for the conditional $p(y|\mathbf{x})$:

$$p(y|\mathbf{x}) = \prod_{k=1}^{K} \theta_k(\mathbf{x})^{[y=k]}, \tag{5.10}$$

where $\theta_k(\mathbf{x})$ is the softmax value for the k-th class, and $[y = k]$ is the Iverson bracket (i.e., $[y = k] = 1$ if y equals k, and 0—otherwise).

Next, we focus on modeling $p(\mathbf{x})$. We can use any marginal model, e.g., we can apply flows and the change of variables formula, namely:

$$p(\mathbf{x}) = \pi\left(\mathbf{z} = f^{-1}(\mathbf{x})\right) |\mathbf{J}_f(\mathbf{x})|^{-1}, \tag{5.11}$$

where $\mathbf{J}_f(\mathbf{x})$ denotes the Jacobian of the transformation (i.e., neural network) f evaluated at $\mathbf{x}$. In the case of the flow, we typically use $\pi(\mathbf{z}) = \mathcal{N}(\mathbf{z}|0, 1)$, i.e., the standard Gaussian distribution.

Plugging these all distributions to the objective of the hybrid modeling $\ell(\mathbf{x}, y; \lambda)$, we get

$$\ell(\mathbf{x}, y; \lambda) = \sum_{k=1}^{K} [y = k] \ln \theta_{k,g,f}(\mathbf{x}) + \lambda \, \mathcal{N}(\mathbf{z} = f^{-1}(\mathbf{x})|0, 1) - \ln |\mathbf{J}_f(\mathbf{x})|,$$

$$(5.12)$$

where we additionally highlight that $\theta_{k,g,f}$ is parameterized by two neural networks: f from the flow and g for the final classification.

Now, if we would like to follow [2], we could pick **coupling layers** as the components of f and, eventually, we would model $p(\mathbf{x})$ using RealNVP or GLOW, for instance. However, we want to be more fancy and we will utilize Integer Discrete Flows (IDFs) [8, 9]. Why? Because we simply can and also IDFs do not require calculating the Jacobian. Besides, we can practice a bit formulating various hybrid models.

Let us quickly recall IDFs. First, they operate on $\mathbb{Z}^D$, i.e., integers. Second, we need to pick an appropriate $\pi(\mathbf{z})$ that in this case could be the **discretized logistic** (DL), $\mathrm{DL}(z|\mu, \nu)$ with mean μ and scale ν. Since the change of variable formula for discrete random variables does not require calculating the Jacobian (remember: no change of volume here!), we can rewrite the hybrid modeling objective as follows:

$$\ell(\mathbf{x}, y; \lambda) = \sum_{k=1}^{K} [y = k] \ln \theta_{k,g,f}(\mathbf{x}) + \lambda \, \mathrm{DL}(\mathbf{z} = f^{-1}(\mathbf{x})|\mu, \nu). \qquad (5.13)$$

That's it! Congratulations, if you have followed all these steps, you have arrived at a new hybrid model that uses IDFs to model the distribution of $\mathbf{x}$. Notice that the classifier takes integers as inputs.

5.4 Code

We have all components to implement our own Hybrid Integer Discrete Flow (HybridIDF)! Below, there is a code with a lot of comments that should help to understand every single line of it. The full code (with auxiliary functions) that you can play with is available at: https://github.com/jmtomczak/intro_dgm.

```
class HybridIDF(nn.Module):
    def __init__(self, netts, classnet, num_flows, alpha=1., D=2)
    :
        super(HybridIDF, self).__init__()

        print('HybridIDF by JT.')

        # Here we use the two options discussed previously: a
    coupling layer or a generalized invertible transformation
        # These formulate the transformation f.
        # NOTE: Please pay attention to a new variable here,
    namely beta. This is the rezero trick used in (van den Berg
    et al., 2020).
```

```
10     if len(netts) == 1:
11         self.t = torch.nn.ModuleList([netts[0]() for _ in
    range(num_flows)])
12         self.idf_git = 1
13         self.beta = nn.Parameter(torch.zeros(len(self.t)))
14
15     elif len(netts) == 4:
16         self.t_a = torch.nn.ModuleList([netts[0]() for _ in
    range(num_flows)])
17         self.t_b = torch.nn.ModuleList([netts[1]() for _ in
    range(num_flows)])
18         self.t_c = torch.nn.ModuleList([netts[2]() for _ in
    range(num_flows)])
19         self.t_d = torch.nn.ModuleList([netts[3]() for _ in
    range(num_flows)])
20         self.idf_git = 4
21         self.beta = nn.Parameter(torch.zeros(len(self.t_a)))
22
23     else:
24         raise ValueError('You can provide either 1 or 4
    translation nets.')
25
26     # This contains extra layers for classification on top of
    z.
27     self.classnet = classnet
28
29     # The number of flows (i.e., f's).
30     self.num_flows = num_flows
31
32     # The rounding operator.
33     self.round = RoundStraightThrough.apply
34
35     # The mean and log-scale for the base distribution pi.
36     self.mean = nn.Parameter(torch.zeros(1, D))
37     self.logscale = nn.Parameter(torch.ones(1, D))
38
39     # The dimensionality of the input.
40     self.D = D
41
42     # Since using ''lambda'' is confusing for Python, we will
    use alpha in the code for lambda in previous equations (not
    confusing at all, right?!)
43     self.alpha = alpha
44
45     # We use the built-in PyTorch loss function. It is for
    educational purposes! Otherwise, we could use the log-
    categorical.
46     self.nll = nn.NLLLoss(reduction='none') #it requires log-
    softmax as input!!
47
48 # The coupling layer as introduced before.
49 # NOTE: We use the rezero trick!
50 def coupling(self, x, index, forward=True):
51
```

```
52        if self.idf_git == 1:
53            (xa, xb) = torch.chunk(x, 2, 1)
54
55            if forward:
56                yb = xb + self.beta[index] * self.round(self.t[
    index](xa))
57            else:
58                yb = xb - self.beta[index] * self.round(self.t[
    index](xa))
59
60            return torch.cat((xa, yb), 1)
61
62        elif self.idf_git == 4:
63            (xa, xb, xc, xd) = torch.chunk(x, 4, 1)
64
65            if forward:
66                ya = xa + self.beta[index] * self.round(self.t_a[
    index](torch.cat((xb, xc, xd), 1)))
67                yb = xb + self.beta[index] * self.round(self.t_b[
    index](torch.cat((ya, xc, xd), 1)))
68                yc = xc + self.beta[index] * self.round(self.t_c[
    index](torch.cat((ya, yb, xd), 1)))
69                yd = xd + self.beta[index] * self.round(self.t_d[
    index](torch.cat((ya, yb, yc), 1)))
70            else:
71                yd = xd - self.beta[index] * self.round(self.t_d[
    index](torch.cat((xa, xb, xc), 1)))
72                yc = xc - self.beta[index] * self.round(self.t_c[
    index](torch.cat((xa, xb, yd), 1)))
73                yb = xb - self.beta[index] * self.round(self.t_b[
    index](torch.cat((xa, yc, yd), 1)))
74                ya = xa - self.beta[index] * self.round(self.t_a[
    index](torch.cat((yb, yc, yd), 1)))
75
76            return torch.cat((ya, yb, yc, yd), 1)
77
78    # The permutation layer.
79    def permute(self, x):
80        return x.flip(1)
81
82    # The flow transformation: forward pass...
83    def f(self, x):
84        z = x
85        for i in range(self.num_flows):
86            z = self.coupling(z, i, forward=True)
87            z = self.permute(z)
88
89        return z
90    # ... and the inverse pass.
91    def f_inv(self, z):
92        x = z
93        for i in reversed(range(self.num_flows)):
94            x = self.permute(x)
95            x = self.coupling(x, i, forward=False)
```

```
96
97      return x
98
99   # A new function: This is used for classification. First we
     predict probabilities, and then pick the most probable value.
100  def classify(self, x):
101      z = self.f(x)
102      y_pred = self.classnet(z) #output: probabilities (i.e.,
     softmax)
103      return torch.argmax(y_pred, dim=1)
104
105  # An auxiliary function: We use it for calculating the
     classification loss, namely the negative log-likelihood for p
     (y|x).
106  # NOTE: We first apply the invertible transformation f.
107  def class_loss(self, x, y):
108      z = self.f(x)
109      y_pred = self.classnet(z) #output: probabilities (i.e.,
     softmax)
110      return self.nll(torch.log(y_pred), y)
111
112  def sample(self, batchSize):
113      # sample z:
114      z = self.prior_sample(batchSize=batchSize, D=self.D)
115      # x = f^-1(z)
116      x = self.f_inv(z)
117      return x.view(batchSize, 1, self.D)
118
119  # The log-probability of the base distribution (a.k.a. prior)
     .
120  def log_prior(self, x):
121      log_p = log_integer_probability(x, self.mean, self.
     logscale)
122      return log_p.sum(1)
123
124  # Sampling from the base distribution.
125  def prior_sample(self, batchSize, D=2):
126      # Sample from logistic
127      y = torch.rand(batchSize, self.D)
128      x = torch.exp(self.logscale) * torch.log(y / (1. - y)) +
     self.mean
129      # And then round it to an integer.
130      return torch.round(x)
131
132  # The forward pass: Now, we use the hybrid model objective!
133  def forward(self, x, y, reduction='avg'):
134      z = self.f(x)
135      y_pred = self.classnet(z) #output: probabilities (i.e.,
     softmax)
136
137      idf_loss = -self.log_prior(z)
138      class_loss = self.nll(torch.log(y_pred), y) #remember to
     use logarithm on top of softmax!
139
```

```
140          if reduction == 'sum':
141              return (class_loss + self.alpha * idf_loss).sum()
142          else:
143              return (class_loss + self.alpha * idf_loss).mean()
```

Listing 5.1 A HybridIDF class

```
1  # The number of invertible transformations
2  num_flows = 2
3
4  # Here, we present only for the option 1 IDF.
5  nett = lambda:nn.Sequential(nn.Linear(D // 2, M), nn.LeakyReLU(),
6                              nn.Linear(M, M), nn.LeakyReLU(),
7                              nn.Linear(M, D // 2))
8  netts = [nett]
9
10 # And a three-layered classifier.
11 classnet = nn.Sequential(nn.Linear(D, M), nn.LeakyReLU(),
12                          nn.Linear(M, M), nn.LeakyReLU(),
13                          nn.Linear(M, K),
14                          nn.Softmax(dim=1))
15
16 # Init HybridIDF
17 model = HybridIDF(netts, classnet, num_flows, D=D, alpha=alpha)
```

Listing 5.2 Examples of neural networks

And we are done, this is all we need to have! After running the code and training the HybridIDFs, we should obtain results similar to those in Fig. 5.4.

5.5 What's Next?

Hybrid VAE The hybrid modeling idea goes beyond using flows for $p(\mathbf{x})$. Instead, e.g., we can pick VAE and then, after applying the variational inference, we get a lower bound to the hybrid modeling objective:

$$\tilde{\ell}(\mathbf{x}, y; \lambda) = \ln p(y|\mathbf{x}) + \lambda \, \mathbb{E}_{\mathbf{z} \sim q(\mathbf{z}|\mathbf{x})} \left[\ln p(\mathbf{x}|\mathbf{z}) + \ln p(\mathbf{z}) - \ln q(\mathbf{z}|\mathbf{x}) \right], \qquad (5.14)$$

where $p(y|\mathbf{x})$ uses the encoder inside, i.e., $q(\mathbf{z}|\mathbf{x})$.

Semi-supervised Hybrid Learning The hybrid modeling perspective is perfectly suited to the semi-supervised scenario. For the labeled data, we can use the objective $\ell(\mathbf{x}, y; \lambda) = \ln p(y|\mathbf{x}) + \lambda \ln p(\mathbf{x})$. However, for the unlabeled data, we can simply consider only the $\ln p(\mathbf{x})$ part. Such approach was used by, e.g., [10] for VAEs.

A very interesting perspective to learning semi-supervised VAE was presented in [11]. The authors end up with an objective that resemblances the hybrid modeling objective but without the cumbersome λ!

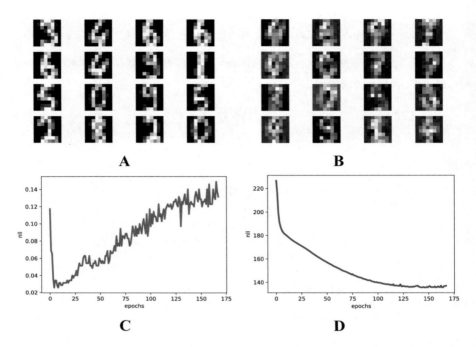

Fig. 5.4 An example of outcomes after the training: (**a**) Randomly selected real images. (**b**) Unconditional generations from the HybridIDF. (**c**) An example of a validation curve for the classification error. (**d**) An example of a validation curve for the negative log-likelihood, i.e., $-\ln p(\mathbf{x})$

The Factor λ As mentioned before, the fudge factor λ could be troublesome. First, it does not follow from a proper probability distribution. Second, it must be tuned that is always extra trouble... But, as mentioned before, [11] showed that we can get rid of λ!

New Parameterizations An interesting open research direction is whether we can get rid of λ by using a different learning algorithm and/or other parameterization (e.g., some peculiar neural networks). I strongly believe it is possible and, one day, we will get there!

Is This a Good Factorization? I am almost sure that some of you wonder whether this factorization of the joint, i.e., $p(\mathbf{x}, y) = p(y|\mathbf{x})\, p(\mathbf{x})$ is indeed better than $p(\mathbf{x}, y) = p(\mathbf{x}|y)\, p(y)$. If I were to sample $\mathbf{x}$ from a specific class y, then the latter is better. However, if you go back to Chap. 1, you will notice that I do not care about *generating*. I prefer to have a good model that will assign proper probabilities to the world. That is why I prefer $p(\mathbf{x}, y) = p(y|\mathbf{x})\, p(\mathbf{x})$.

References

1. Guillaume Bouchard and Bill Triggs. The tradeoff between generative and discriminative classifiers. In *16th IASC International Symposium on Computational Statistics (COMPSTAT'04)*, pages 721–728, 2004.
2. Eric Nalisnick, Akihiro Matsukawa, Yee Whye Teh, Dilan Gorur, and Balaji Lakshminarayanan. Hybrid models with deep and invertible features. In *International Conference on Machine Learning*, pages 4723–4732. PMLR, 2019.
3. Diederik P Kingma, Danilo J Rezende, Shakir Mohamed, and Max Welling. Semi-supervised learning with deep generative models. In *Proceedings of the 27th International Conference on Neural Information Processing Systems*, pages 3581–3589, 2014.
4. Sergey Tulyakov, Andrew Fitzgibbon, and Sebastian Nowozin. Hybrid VAE: Improving deep generative models using partial observations. *arXiv preprint arXiv:1711.11566*, 2017.
5. Diederik P Kingma and Prafulla Dhariwal. Glow: generative flow with invertible 1×1 convolutions. In *Proceedings of the 32nd International Conference on Neural Information Processing Systems*, pages 10236–10245, 2018.
6. Ricky TQ Chen, Jens Behrmann, David Duvenaud, and Jörn-Henrik Jacobsen. Residual flows for invertible generative modeling. *arXiv preprint arXiv:1906.02735*, 2019.
7. Yura Perugachi-Diaz, Jakub M Tomczak, and Sandjai Bhulai. Invertible DenseNets with concatenated LipSwish. *Advances in Neural Information Processing Systems*, 2021.
8. Rianne van den Berg, Alexey A Gritsenko, Mostafa Dehghani, Casper Kaae Sønderby, and Tim Salimans. Idf++: Analyzing and improving integer discrete flows for lossless compression. *arXiv e-prints*, pages arXiv–2006, 2020.
9. Emiel Hoogeboom, Jorn WT Peters, Rianne van den Berg, and Max Welling. Integer discrete flows and lossless compression. *arXiv preprint arXiv:1905.07376*, 2019.
10. Maximilian Ilse, Jakub M Tomczak, Christos Louizos, and Max Welling. DIVA: Domain invariant variational autoencoders. In *Medical Imaging with Deep Learning*, pages 322–348. PMLR, 2020.
11. Tom Joy, Sebastian M Schmon, Philip HS Torr, N Siddharth, and Tom Rainforth. Rethinking semi-supervised learning in VAEs. *arXiv preprint arXiv:2006.10102*, 2020.

Chapter 6
Energy-Based Models

6.1 Introduction

So far, we have discussed various deep generative models for modeling the marginal distribution over observable variables (e.g., images), $p(\mathbf{x})$, such as, autoregressive models (ARMs), flow-based models (flows, for short), Variational Auto-Encoders (VAEs), and hierarchical models like hierarchical VAEs and diffusion-based deep generative models (DDGMs). However, from the very beginning, we advocate for using deep generative modeling in the context of finding the joint distribution over observables and decision variables that is factorized as $p(\mathbf{x}, y) = p(y|\mathbf{x})p(\mathbf{x})$. After taking the logarithm of the joint we obtain two additive components: $\ln p(\mathbf{x}, y) = \ln p(y|\mathbf{x}) + \ln p(\mathbf{x})$. We outlined how such a joint model could be formulated and trained in the hybrid modeling setting (see Chap. 5). The drawback of hybrid modeling though is the necessity of weighting both distributions, i.e., $\ell(\mathbf{x}, y\lambda) = \ln p(y|\mathbf{x}) + \lambda \ln p(\mathbf{x})$, and for $\lambda \neq 1$ this objective does not correspond to the log-likelihood of the joint distribution. The question is whether it is possible to formulate a model to learn with $\lambda = 1$. Here, we are going to discuss a potential solution to this problem using probabilistic **energy-based models** (EBMs) [1].

The history of EBMs is long and dates back to 80 of the previous century when models dubbed **Boltzmann Machines** were proposed [2, 3]. Interestingly, the idea behind Boltzmann machines is taken from statistical physics and was formulated by cognitive scientists. A nice mix-up, isn't it? In a nutshell, instead of proposing a specific distribution like Gaussian or Bernoulli, we can define an **energy function**, $E(\mathbf{x})$, that assigns a value (*energy*) to a given state. There are no restrictions on the energy function so you can already think of parameterizing it with neural networks. Then, the probability distribution could be obtained by transforming the energy to the unnormalized probability $e^{-E(\mathbf{x})}$ and normalizing it by $Z = \sum_{\mathbf{x}} e^{-E(\mathbf{x})}$ (a.k.a. a *partition function*) that yields the Boltzmann (also called Gibbs) distribution:

© The Author(s), under exclusive license to Springer Nature Switzerland AG 2022
J. M. Tomczak, *Deep Generative Modeling*,
https://doi.org/10.1007/978-3-030-93158-2_6

$$p(\mathbf{x}) = \frac{e^{-E(\mathbf{x})}}{Z}. \tag{6.1}$$

If we consider continuous random variables, then the sum sign should be replaced by the integral. In physics, the energy is scaled by an inverse of temperature [4], however, we skip it to keep the notation uncluttered. Understanding how the Boltzmann distribution works is relatively simple. Imagine a grid 5-by-5. Then, assign some value (energy) to each of the 25 points where a larger value means that a point has higher energy. Exponentiating the energy ensures that we do not obtain negative values. To calculate the probability for each point, we need to divide all exponentiated energies by their sum, in the same way how we do it for calculating softmax. In the case of continuous random variables, we must normalize by calculating an integral (i.e., a sum over all infinitesimal regions). For instance, the Gaussian distribution could be also expressed as the Boltzmann distribution with an analytically tractable partition function and the energy function of the following form:

$$E(x; \mu, \sigma^2) = \frac{1}{2\sigma^2}(x - \mu)^2, \tag{6.2}$$

that yields

$$p(x) = \frac{e^{-E(x)}}{\int e^{-E(x)} \mathrm{d}x} \tag{6.3}$$

$$= \frac{e^{\frac{1}{2\sigma^2}(x-\mu)^2}}{\sqrt{2\pi\sigma^2}}. \tag{6.4}$$

In practice, most energy functions do not result in a nicely computable partition function. And, typically, the partition function is the key element that is problematic in learning energy-based models. The second problem is that, in general, it is hard to sample from such models. Why? Well, we know the probability for each point but there is no generative process like in ARMs, flows, or VAEs. It is unclear how to start and what is the graphical model for an EBM. We can think of the EBM as a box that for a given $\mathbf{x}$ can tell us the (unnormalized) probability of that point. Notice that the energy function does not distinguish variables in any way, it does not care about any structure in $\mathbf{x}$. It says: Give me $\mathbf{x}$ and I will return the value. That's it! In other words, the energy function defines mountains and valleys over the space of random variables.

A curious reader (yes, I am referring to you!) may ask why we want to bother with EBMs. Previously discussed models are at least tractable and comprehensible in the sense that some stochastic dependencies are defined. Now we suddenly invert the logic and say that we do not care about modeling the structure and instead we want to model an energy function that returns unnormalized probabilities. Is it beneficial? Yes, for at least three reasons. First, in principle, the energy function is unconstrained, it could be any function! Yes, you have probably guessed already, it could be a neural network! Second, notice that the energy function could be multimodal without being defined as such (i.e., opposing to a mixture distribution). Third, there is no difference if we define it over discrete or continuous variables. I hope you see now that EBMs have a lot of advantages! They possess a lot of deficiencies too but hey, let us stick to the positive aspects before we start, ok?

6.2 Model Formulation

As mentioned earlier, we formulate an energy function with some parameters θ over observable and decision random variables, $E(\mathbf{x}, y; \theta)$, that assigns a value (an energy) to a pair $(\mathbf{x}, y)$ where $\mathbf{x} \in \mathbb{R}^D$ and $y \in \{0, 1, \ldots, K - 1\}$. Let $E(\mathbf{x}, y; \theta)$ be parameterized by a neural network $NN_\theta(\mathbf{x})$ that returns K values: $NN_\theta : \mathbb{R}^D \to \mathbb{R}^K$. In other words, we can define the energy as follows:

$$E(\mathbf{x}, y; \theta) = -NN_\theta(\mathbf{x})[y], \tag{6.5}$$

where we indicate by $[y]$ the specific output of the neural net $NN_\theta(\mathbf{x})$. Then, the joint probability distribution is defined as the Boltzmann distribution:

$$p_\theta(\mathbf{x}, y) = \frac{\exp\{NN_\theta(\mathbf{x})[y]\}}{\sum_{\mathbf{x}, y} \exp\{NN_\theta(\mathbf{x})[y]\}} \tag{6.6}$$

$$= \frac{\exp\{NN_\theta(\mathbf{x})[y]\}}{Z_\theta}, \tag{6.7}$$

where we define the partition function as $Z_\theta = \sum_{\mathbf{x}, y} \exp\{NN_\theta(\mathbf{x})[y]\}$.

Since we have the joint distribution, we can calculate the marginal distribution and the conditional distribution. First, let us take a look at the marginal $p(\mathbf{x})$. Applying the sum rule to the joint distribution yields:

$$p_\theta(\mathbf{x}) = \sum_y p_\theta(\mathbf{x}, y) \tag{6.8}$$

$$= \frac{\sum_y \exp\{NN_\theta(\mathbf{x})[y]\}}{\sum_{\mathbf{x}, y} \exp\{NN_\theta(\mathbf{x})\}[y]} \tag{6.9}$$

$$= \frac{\sum_y \exp\{NN_\theta(\mathbf{x})[y]\}}{Z_\theta}. \tag{6.10}$$

Let us notice that we can express this distribution differently. First, we can rewrite the numerator in the following manner:

$$\sum_y \exp\{NN_\theta(\mathbf{x})[y]\} = \exp\left\{\log\left(\sum_y \exp\{NN_\theta(\mathbf{x})[y]\}\right)\right\} \qquad (6.11)$$

$$= \exp\left\{\text{LogSumExp}_y\{NN_\theta(\mathbf{x})[y]\}\right\} \qquad (6.12)$$

where we define $\text{LogSumExp}_y\{f(y)\} = \ln\sum_y \exp\{f(y)\}$. In other words, we can say that the energy function of the marginal distribution is expressed as $-\text{LogSumExp}_y\{NN_\theta(\mathbf{x})[y]\}$. Then, the marginal distribution could be defined as follows:

$$p_\theta(\mathbf{x}) = \frac{\exp\left\{\text{LogSumExp}_y\{NN_\theta(\mathbf{x})[y]\}\right\}}{Z_\theta}. \qquad (6.13)$$

Now, we can calculate the conditional distribution $p_\theta(y|\mathbf{x})$. We know that $p_\theta(\mathbf{x}, y) = p_\theta(y|\mathbf{x})\, p_\theta(\mathbf{x})$ thus:

$$p_\theta(y|\mathbf{x}) = \frac{p_\theta(\mathbf{x}, y)}{p_\theta(\mathbf{x})} \qquad (6.14)$$

$$= \frac{\frac{\exp\{NN_\theta(\mathbf{x})[y]\}}{Z_\theta}}{\frac{\sum_y \exp\{NN_\theta(\mathbf{x})[y]\}}{Z_\theta}} \qquad (6.15)$$

$$= \frac{\exp\{NN_\theta(\mathbf{x})[y]\}}{\sum_y \exp\{NN_\theta(\mathbf{x})[y]\}}. \qquad (6.16)$$

The last line should resemble something, do you see it? Yes, you are right, it is the **softmax** function! We have shown that the energy-based model could be used either as a classifier or as a marginal distribution. And it is enough to define a single neural network for that! Isn't it beautiful? The same observations were made in [5] that any classifier could be seen as an energy-based model.

Interestingly, the logarithm of the joint distribution is the following:

$$\ln p_\theta(\mathbf{x}, y) = \ln p_\theta(y|\mathbf{x}) + \ln p_\theta(\mathbf{x}) \qquad (6.17)$$

$$= \ln \frac{\exp\{NN_\theta(\mathbf{x})[y]\}}{\sum_y \exp\{NN_\theta(\mathbf{x})[y]\}} + \ln \frac{\sum_y \exp\{NN_\theta(\mathbf{x})[y]\}}{Z_\theta} \qquad (6.18)$$

$$= \ln \text{softmax}\{NN_\theta(\mathbf{x})[y]\} + \left(\text{LogSumExp}_y\{NN_\theta(\mathbf{x})[y]\} - \ln Z_\theta\right), \qquad (6.19)$$

where we define $\text{LogSumExp}_y\{f(y)\} = \ln\sum_y \exp\{f(y)\}$. We clearly see that the model requires a shared neural network that is used for calculating both

Fig. 6.1 A schematic
representation of an EBM.
We denote the output of
LogSumExp$_y$ by $\tilde{p}_\theta(\mathbf{x})$ to
highlight that it is the
unnormalized distribution
since calculating the partition
function is troublesome

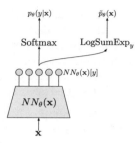

distributions. To obtain a specific distribution, we pick the final activation function.
The model is schematically presented in Fig. 6.1.

6.3   Training

We have a single neural network to train and the training objective is the logarithm
of the joint distribution. Since the training objective is a sum of the logarithm
of the conditional $p_\theta(y|\mathbf{x})$ and the logarithm the marginal $p_\theta(\mathbf{x})$, calculating the
gradient with respect to the parameters θ requires taking the gradient of each of the
component separately. We know that there is no problem with learning a classifier
so let us take a closer look at the second component, namely:

$$\nabla_\theta \ln p_\theta(\mathbf{x}) = \nabla_\theta \text{LogSumExp}_y\{NN_\theta(\mathbf{x})[y]\} - \nabla_\theta \ln Z_\theta \tag{6.20}$$

$$= \nabla_\theta \text{LogSumExp}_y\{NN_\theta(\mathbf{x})[y]\}+$$
$$-\nabla_\theta \ln \sum_{\mathbf{x}} \exp\left\{\text{LogSumExp}_y\{NN_\theta(\mathbf{x})[y]\}\right\} \tag{6.21}$$

$$= \nabla_\theta \text{LogSumExp}_y\{NN_\theta(\mathbf{x})[y]\}+$$
$$-\sum_{\mathbf{x}'} \frac{\exp\left\{\text{LogSumExp}_y\{NN_\theta(\mathbf{x}')[y]\}\right\}}{\sum_{\mathbf{x}'',y''}\exp\{NN_\theta(\mathbf{x}'')[y'']\}} \nabla_\theta \text{LogSumExp}_y\{NN_\theta(\mathbf{x}')[y]\}$$
$$\tag{6.22}$$

$$= \nabla_\theta \text{LogSumExp}_y\{NN_\theta(\mathbf{x})[y]\}+$$
$$-\mathbb{E}_{\mathbf{x}'\sim p_\theta(\mathbf{x})}\left[\nabla_\theta \text{LogSumExp}_y\{NN_\theta(\mathbf{x}')[y]\}\right]. \tag{6.23}$$

We can decipher what has just happened here. The gradient of the first part,
$\nabla_\theta \text{LogSumExp}_y\{NN_\theta(\mathbf{x})[y]\}$, is calculated for a given datapoint $\mathbf{x}$. The log-sum-
exp function is differentiable, so we can apply autograd tools. However, the second
part, $\mathbb{E}_{\mathbf{x}'\sim p_\theta(\mathbf{x})}\left[\nabla_\theta \text{LogSumExp}_y\{NN_\theta(\mathbf{x}')[y]\}\right]$, is a totally different story for two
reasons:

- First, the gradient of the logarithm of the partition function turns into the expected value over $\mathbf{x}$ distributed according to the model! That is really a problem because the expected value cannot be analytically calculated and sampling from the marginal distribution $p_\theta(\mathbf{x})$ is non-trivial.
- Second, we need to calculate the expected value of the log-sum-exp of $NN_\theta(\mathbf{x})$. That is good news because we can do it using automatic differentiation tools.

Thus, the only problem lies in the expected value. Typically, it is approximated by Monte Carlo samples, however, it is not clear how to sample effectively and efficiently from an EBM. Grathwohl et al. [5] proposes to use the Langevin dynamics [6] that is an MCMC method. The Langevin dynamics in our case starts with a randomly initialized $\mathbf{x}_0$ and then uses the information about the landscape of the energy function (i.e., the gradient) to seek for new $\mathbf{x}$, that is

$$\mathbf{x}_{t+1} = \mathbf{x}_t + \alpha \nabla_{\mathbf{x}_t} \text{LogSumExp}_y \{NN_\theta(\mathbf{x})[y]\} + \sigma \cdot \epsilon, \tag{6.24}$$

where $\alpha > 0$, $\sigma > 0$, and $\epsilon \sim \mathcal{N}(0, I)$. The Langevin dynamics could be seen as the stochastic gradient descent in the observable space with a small Gaussian noise added at each step. Once we apply this procedure for η steps, we can approximate the gradient as follows:

$$\nabla_\theta \ln p_\theta(\mathbf{x}) \approx \nabla_\theta \text{LogSumExp}_y \{NN_\theta(\mathbf{x})[y]\} - \nabla_\theta \text{LogSumExp}_y \{NN_\theta(\mathbf{x}_\eta)[y]\}, \tag{6.25}$$

where $\mathbf{x}_\eta$ denotes the last step of the Langevin dynamics procedure.

We are ready to put it all together! Please remember that our training objective is the following:

$$\ln p_\theta(\mathbf{x}, y) = \ln \text{softmax}\{NN_\theta(\mathbf{x})[y]\} + \left(\text{LogSumExp}_y\{NN_\theta(\mathbf{x})[y]\} - \ln Z_\theta\right), \tag{6.26}$$

where the first part is for learning a classifier, and the second part is for learning a generator (so to speak). As a result, we can say we have a sum of two objectives for a fully shared model. The gradient with respect to the parameters is the following:

$$\begin{aligned} \nabla_\theta \ln p_\theta(\mathbf{x}, y) = &\nabla_\theta \ln \text{softmax}\{NN_\theta(\mathbf{x})[y]\} + \\ &+ \nabla_\theta \text{LogSumExp}_y\{NN_\theta(\mathbf{x})[y]\} + \\ &- \mathbb{E}_{\mathbf{x}' \sim p_\theta(\mathbf{x})} \left[\nabla_\theta \text{LogSumExp}_y\{NN_\theta(\mathbf{x}')[y]\}\right]. \end{aligned} \tag{6.27}$$

The last two components come from calculating the gradient of the marginal distribution. Remember that the problematic part is only the last component! We will approximate this part using the Langevine dynamics (i.e., a sampling procedure) and a single sample. The final training procedure is the following:

1. Sample $\mathbf{x}_n$ and y_n from a dataset.

2. Calculate $NN_\theta(\mathbf{x}_n)[y]$.
3. Initialize $\mathbf{x}_0$ using, e.g., a uniform distribution.
4. Run the Langevin dynamics for η steps:

$$\mathbf{x}_{t+1} = \mathbf{x}_t + \alpha \nabla_{\mathbf{x}_t} \text{LogSumExp}_y \{NN_\theta(\mathbf{x})[y]\} + \sigma \cdot \epsilon. \tag{6.28}$$

5. Calculate the objective:

$$L_{clf}(\theta) = \sum_y \mathbf{1}[y = y_n] \, \theta_y \ln \{NN_\theta(\mathbf{x}_n)[y]\} \tag{6.29}$$

$$L_{gen}(\theta) = \text{LogSumExp}_y\{NN_\theta(\mathbf{x})[y]\} - \text{LogSumExp}_y\{NN_\theta(\mathbf{x}_\eta)[y]\} \tag{6.30}$$

$$L(\theta) = L_{clf}(\theta) + L_{gen}(\theta). \tag{6.31}$$

6. Apply the autograd tool to calculate gradients $\nabla_\theta L(\theta)$ and update the neural network.

Notice that $L_{clf}(\theta)$ is nothing else than the cross-entropy loss, and $L_{gen}(\theta)$ is a (crude) approximation to the log-marginal distribution over $\mathbf{x}$'s.

6.4   Code

What do we need to code then? First, we must specify the neural network that defines the energy function. (let us call it the *energy net*.) Classifying using the energy net is rather straightforward. The main problematic part is sampling from the model using the Langevin dynamics. Fortunately, the autograd tools allow us to easily access the gradient with respect $\mathbf{x}$! In fact, it is a single line in the code below. Then we require writing a loop to run the Langevin dynamics for η iterations with the steps size α and the noise level equal σ. In the code, we assume the data are normalized and scaled to $[-1, 1]$ similarly to [5].

```
1  class EBM(nn.Module):
2      def __init__(self, energy_net, alpha, sigma, ld_steps, D):
3          super(EBM, self).__init__()
4
5          print('EBM by JT.')
6
7          # the neural net used by the EBM
8          self.energy_net = energy_net
9
10          # the loss for classification
11          self.nll = nn.NLLLoss(reduction='none')  # it requires
    log-softmax as input!!
12
13          # hyperparams
```

```
14          self.D = D
15
16          self.sigma = sigma
17
18          self.alpha = torch.FloatTensor([alpha])
19
20          self.ld_steps = ld_steps
21
22      def classify(self, x):
23          f_xy = self.energy_net(x)
24          y_pred = torch.softmax(f_xy, 1)
25          return torch.argmax(y_pred, dim=1)
26
27      def class_loss(self, f_xy, y):
28          # - calculate logits (for classification)
29          y_pred = torch.softmax(f_xy, 1)
30
31          return self.nll(torch.log(y_pred), y)
32
33      def gen_loss(self, x, f_xy):
34          # - sample using Langevin dynamics
35          x_sample = self.sample(x=None, batch_size=x.shape[0])
36
37          # - calculate f(x_sample)[y]
38          f_x_sample_y = self.energy_net(x_sample)
39
40          return -(torch.logsumexp(f_xy, 1) - torch.logsumexp(
    f_x_sample_y, 1))
41
42      def forward(self, x, y, reduction='avg'):
43          # =====
44          # forward pass through the network
45          # - calculate f(x)[y]
46          f_xy = self.energy_net(x)
47
48          # =====
49          # discriminative part
50          # - calculate the discriminative loss: the cross-entropy
51          L_clf = self.class_loss(f_xy, y)
52
53          # =====
54          # generative part
55          # - calculate the generative loss: E(x) - E(x_sample)
56          L_gen = self.gen_loss(x, f_xy)
57
58          # =====
59          # Final objective
60          if reduction == 'sum':
61              loss = (L_clf + L_gen).sum()
62          else:
63              loss = (L_clf + L_gen).mean()
64
65          return loss
66
```

```
67    def energy_gradient(self, x):
68        self.energy_net.eval()
69
70        # copy original data that doesn't require grads!
71        x_i = torch.FloatTensor(x.data)
72        x_i.requires_grad = True   # WE MUST ADD IT, otherwise
      autograd won't work
73
74        # calculate the gradient
75        x_i_grad = torch.autograd.grad(torch.logsumexp(self.
      energy_net(x_i), 1).sum(), [x_i], retain_graph=True)[0]
76
77        self.energy_net.train()
78
79        return x_i_grad
80
81    def langevine_dynamics_step(self, x_old, alpha):
82        # Calculate gradient wrt x_old
83        grad_energy = self.energy_gradient(x_old)
84        # Sample eta ~ Normal(0, alpha)
85        epsilon = torch.randn_like(grad_energy) * self.sigma
86
87        # New sample
88        x_new = x_old + alpha * grad_energy + epsilon
89
90        return x_new
91
92    def sample(self, batch_size=64, x=None):
93        # - 1) Sample from uniform
94        x_sample = 2. * torch.rand([batch_size, self.D]) - 1.
95
96        # - 2) run Langevin Dynamics
97        for i in range(self.ld_steps):
98            x_sample = self.langevine_dynamics_step(x_sample,
      alpha=self.alpha)
99
100       return x_sample
```

Listing 6.1 A EBM class

And we are done, this is all we need to have! After running the code (take a look at: https://github.com/jmtomczak/intro_dgm) and training the EBM, we should obtain results similar to those in Fig. 6.2.

6.5 Restricted Boltzmann Machines

The idea of defining a model through the energy function is the foundation of a broad family of Boltzmann machines (BMs) [2, 7]. The Boltzmann machines define an energy function as follows:

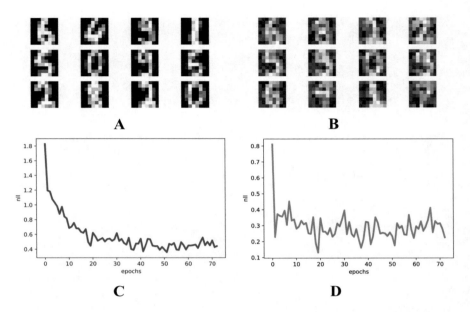

Fig. 6.2 An example of outcomes after the training: (**a**) Randomly selected real images. (**b**) Unconditional generations from the EBM after applying $\eta = 20$ steps of the Langevin dynamics. (**c**) An example of a validation curve of the objective $(L_{clf} + L_{gen})$. (**d**) An example of a validation curve of the generative objective L_{gen}

$$E(\mathbf{x}; \theta) = -\left(\mathbf{x}^\top \mathbf{W} \mathbf{x} + \mathbf{b}^\top \mathbf{x}\right), \tag{6.32}$$

where $\theta = \{\mathbf{W}, \mathbf{b}\}$ and $\mathbf{W}$ is the weight matrix and $\mathbf{b}$ is the bias vector (bias weights), which is the same as that in Hopfield networks and Ising models. The problem with BM is that they are hard to train (due to the partition function). However, we can alleviate the problem by introducing latent variables and restricting connections among observables.

Restricting BMs

Let us consider a BM that consists of binary observable variables $\mathbf{x} \in \{0, 1\}^D$ and binary latent (hidden) variables $\mathbf{z} \in \{0, 1\}^M$. The relationships among variables are specified through the following *energy function*:

$$E(\mathbf{x}, \mathbf{z}; \theta) = -\mathbf{x}^\top \mathbf{W} \mathbf{z} - \mathbf{b}^\top \mathbf{x} - \mathbf{c}^\top \mathbf{z}, \tag{6.33}$$

where $\theta = \{\mathbf{W}, \mathbf{b}, \mathbf{c}\}$ is a set of parameters, $\mathbf{W} \in \mathbb{R}^{D \times M}$, $\mathbf{b} \in \mathbb{R}^D$, and $\mathbf{c} \in \mathbb{R}^M$ are, respectively, weights, observable biases, and hidden biases. For the energy function in Eq. (6.33), RBM is defined by the *Gibbs distribution*:

$$p(\mathbf{x}, \mathbf{z}|\theta) = \frac{1}{Z_\theta} \exp\big(- E(\mathbf{x}, \mathbf{z}; \theta)\big), \tag{6.34}$$

where

$$Z_\theta = \sum_{\mathbf{x}} \sum_{\mathbf{z}} \exp\big(- E(\mathbf{x}, \mathbf{z}; \theta)\big) \tag{6.35}$$

is the *partition function*. The marginal probability over observables (the likelihood of an observation) is

$$p(\mathbf{x}|\theta) = \frac{1}{Z_\theta} \exp\big(- F(\mathbf{x}; \theta)\big), \tag{6.36}$$

where $F(\cdot)$ is the *free energy*:[1]

$$F(\mathbf{x}; \theta) = -\mathbf{b}^\top \mathbf{x} - \sum_j \log \Big(1 + \exp(\mathbf{c}_j + (\mathbf{W}_{\cdot j})^\top \mathbf{x})\Big). \tag{6.37}$$

The presented model is called a Restricted Boltzmann Machine (RBM). It possesses the very useful property that the conditional distribution over the hidden variables factorizes given the observable variables and *vice versa*, which yields the following:

$$p(\mathbf{z}_m = 1|\mathbf{x}, \theta) = \mathrm{sigm}\big(\mathbf{c}_m + (\mathbf{W}_{\cdot m})^\top \mathbf{x}\big), \tag{6.38}$$

$$p(\mathbf{x}_d = 1|\mathbf{z}, \theta) = \mathrm{sigm}(\mathbf{b}_d + \mathbf{W}_{d\cdot}\mathbf{z}). \tag{6.39}$$

Learning RBMs

For given data $\mathcal{D} = \{\mathbf{x}_n\}_{n=1}^N$, we can train an RBM using the maximum likelihood approach that seeks the maximum of the log-likelihood function:

$$\ell(\theta) = \frac{1}{N} \sum_{\mathbf{x}_n \in \mathcal{D}} \log p(\mathbf{x}_n|\theta). \tag{6.40}$$

The gradient of the learning objective $\ell(\theta)$ with respect to θ takes the following form:

$$\nabla_\theta \ell(\theta) = -\frac{1}{N} \sum_{n=1}^N \Big(\nabla_\theta F(\mathbf{x}_n; \theta) - \sum_{\hat{\mathbf{x}}} p(\hat{\mathbf{x}}|\theta) \nabla_\theta F(\hat{\mathbf{x}}; \theta)\Big). \tag{6.41}$$

---

[1] We use the following notation: for given matrix $\mathbf{A}$, $\mathbf{A}_{ij}$ is its element at location (i, j), $\mathbf{A}_{\cdot j}$ denotes its jth column , $\mathbf{A}_{i\cdot}$ denotes its ith row, and for given vector $\mathbf{a}$, $\mathbf{a}_i$ is its ith element.

In general, the gradient in Eq. (6.41) cannot be computed analytically because the second term requires summing over all configurations of observables. One way to sidestep this issue is the standard stochastic approximation of replacing the expectation under $p(\mathbf{x}|\theta)$ by a sum over S samples $\{\hat{\mathbf{x}}_1, \ldots, \hat{\mathbf{x}}_S\}$ drawn according to $p(\mathbf{x}|\theta)$ [8]:

$$\nabla_\theta \ell(\theta) \approx -\Big(\frac{1}{N} \sum_{n=1}^{N} \nabla_\theta F(\mathbf{x}_n; \theta) - \frac{1}{S} \sum_{s=1}^{S} \nabla_\theta F(\hat{\mathbf{x}}_s; \theta)\Big). \tag{6.42}$$

A different approach, *contrastive divergence*, approximates the expectation under $p(\mathbf{x}|\theta)$ in Eq. (6.41) by a sum over samples $\bar{\mathbf{x}}_n$ drawn from a distribution obtained by applying K steps of the block-Gibbs sampling procedure:

$$\nabla_\theta \ell(\theta) \approx -\frac{1}{N} \sum_{n=1}^{N} \Big(\nabla_\theta F(\mathbf{x}_n; \theta) - \nabla_\theta F(\bar{\mathbf{x}}_n; \theta)\Big). \tag{6.43}$$

The original CD [9] used K steps of the Gibbs chain, starting from each datapoint $\mathbf{x}_n$ to obtain a sample $\bar{\mathbf{x}}_n$ and is restarted after every parameter update. An alternative approach, *Persistent Contrastive Divergence* (PCD), does not restart the chain after each update; this typically results in a slower convergence rate but eventually better performance [10].

Defining Higher-Order Relationships Through the Energy Function
The energy function is an interesting concept because it allows modeling higher-order dependencies among variables. For instance, the binary RBM could be extended to third-order multiplicative interactions by introducing two kinds of hidden variables, i.e., subspace units and gate units. The subspace units are hidden variables that reflect variations of a feature, and, thus, they are more robust to invariances. The gate units are responsible for activating the subspace units and they can be seen as pooling features composed of the subspace features.

Let us consider the following random variables: $\mathbf{x} \in \{0, 1\}^D$, $\mathbf{h} \in \{0, 1\}^M$, $\mathbf{S} \in \{0, 1\}^{M \times K}$. We are interested in the situation where there are three variables connected, namely one observable x_i and two types of hidden binary units, a gate unit h_j and a subspace unit s_{jk}. Each gate unit is associated with a group of subspace hidden units. The energy function of a joint configuration is then defined as follows:[2]

$$E(\mathbf{x}, \mathbf{h}, \mathbf{S}; \boldsymbol{\theta}) = -\sum_{i=1}^{D} \sum_{j=1}^{M} \sum_{k=1}^{K} W_{ijk} x_i h_j s_{jk} - \sum_{i=1}^{D} b_i x_i - \sum_{j=1}^{M} c_j h_j - \sum_{j=1}^{M} h_j \sum_{k=1}^{K} D_{jk} s_{jk}, \tag{6.44}$$

---

[2] Unlike in other cases, we use sums instead of matrix products because now we have third-order multiplications that would complicate the notation.

where the parameters are $\theta = \{\mathbf{W}, \mathbf{b}, \mathbf{c}, \mathbf{D}\}$, where $\mathbf{W} \in \mathbb{R}^{D \times M \times K}$, $\mathbf{b} \in \mathbb{R}^{D}$, $\mathbf{c} \in \mathbb{R}^{M}$, and $\mathbf{D} \in \mathbb{R}^{M \times K}$.

The Gibbs distribution with the energy function in (6.44) is called *subspace Restricted Boltzmann Machine* (subspaceRBM) [11]. For the subspaceRBM the following conditional dependencies hold true:[3]

$$p(x_i = 1|\mathbf{h}, \mathbf{S}) = \mathrm{sigm}\Big(\sum_j \sum_k W_{ijk} h_j s_{jk} + b_i\Big), \tag{6.45}$$

$$p(s_{jk} = 1|\mathbf{x}, h_j) = \mathrm{sigm}\Big(\sum_i W_{ijk} x_i h_j + h_j D_{jk}\Big), \tag{6.46}$$

$$p(h_j = 1|\mathbf{x}) = \mathrm{sigm}\Big(-K\log 2 + c_j + \sum_{k=1}^{K} \mathrm{softplus}\Big(\sum_i W_{ijk} x_i + D_{jk}\Big)\Big), \tag{6.47}$$

which can be straightforwardly used in formulating a contrastive divergence-like learning algorithm. Notice that in (6.47) a term $-K\log 2$ imposes a natural penalty of the hidden unit activation which is linear to the number of subspace hidden variables. Therefore, the gate unit is inactive unless the sum of softplus of total input exceeds the penalty term and the bias term.

The example of the subspaceRBM shows that the energy function is handy and allows the modeling of various stochastic relations. The subspaceRBM was used to model invariance features but other modifications of the energy function in RBMs could be formulated to allow training spatial transformations [12] or spike-and-slab features [13].

6.6 Final Remarks

The paper of [5] is definitely a milestone in the EBM literature because it shows that we can use *any* energy function parameterized by a neural network. However, to get to that point there was a lot of work on the energy-based models.

Restricted Boltzmann Machines RBMs possess a couple of useful traits. First, the bipartite structure helps training that could be further used in formulating an efficient training procedure for RBMs called contrastive divergence [9] that takes advantage of block-Gibbs sampling. As mentioned earlier, a chain is initialized either at a random point or a sample of latents and then, conditionally, the other set of variables are trained. Similar to the ping-pong game, we sample some variables given the others until convergence or until we decide to stop. Second, the distribution

---

[3] $\mathrm{softplus}(a) = \log\big(1 + \exp(a)\big)$.

over latent variables could be calculated analytically. Moreover, it could be seen as being parameterized by logistic regression. That is an interesting fact that the sigmoid function arises naturally from the definition of the energy function! Third, the restrictions among connections show that we can still build powerful models that are (partially) analytically tractable. This opened a new research direction that aimed for formulating models with more sophisticated structures like spike-and-slab RBMs [13] and higher-order RBMs [11, 12], or RBMs for categorical observables [14] or real-valued observables [15]. Moreover, RBMs could be also modified to handle temporal data [16] that could be applied to, e.g., human motion tracking [17]. The precursor of the idea presented in [5] was the work on classification RBMs [18, 19]. The training of RBMs is based on the MCMC techniques, e.g., the contrastive divergence algorithm. However, RBMs could be trained to achieve specific features by regularization [20] or other learning algorithms like the Perturb-and-MAP method [21, 22], minimum probability flow [23], or other algorithms [8, 24].

Deep Boltzmann Machines A natural extension of BMs are models with a deep architecture or hierarchical BMs. As indicated by many, the idea of hierarchical models plays a crucial role in AI [25], therefore, there are many extensions of BMs with hierarchical (deep) architectures [15, 26, 27].

Training of deep BMs is even more challenging due to the complexity of the partition function [28]. One of the main approaches to the training of deep BMs relies on treating each pair of consecutive layers as an RBM and training them layer by layer where the layer at the lower layer is treated as observed [27]. This procedure was successfully applied in the seminal paper on unsupervised pre-training of neural nets [29].

Approximating the Partition Function The crucial quantity in the EBMs is the partition function because it allows calculating the Boltzmann distribution. Unfortunately, summing over all values of random variables is computationally infeasible. However, we can use one of the available approximation techniques:

- *Variational methods*: There are a few variational methods that lower bound the log-partition function using the Bethe approximation [30, 31] or upper-bound the log-partition function using a tree-reweighted sum-product algorithm [32]. *Perturb-and-MAP methods*: an alternative approach is to relate the partition function to the max-statistics of random variables and apply the Perturb-and-MAP method [33].
- *Stochastic approximations*: a different approach, probably the most straightforward, is to utilize a sampling procedure. One widely used technique is the Annealed Importance Sampling [28].

Some of the approximations are useful for specific BMs, e.g., BMs with binary variables, BMs with a specific structure. In general, however, approximating the partition function remains an open question and is the main road-blocker for using EMBs in practice and on a large scale.

EBMs Are the Future?
There is definitely a lot of potential in EBMs for at least two reasons:

1. They do not require using any fudge factor to balance the classification loss and the generative loss like in the hybrid modeling approach.
2. The results obtained by Grathwohl et al.[5] clearly indicate that EBMs can achieve the SOTA classification error, synthesize images of high fidelity and be of great use for the out-of-distribution selection.

However, there is one main problem that has not been yet solved: The calculation of $p(\mathbf{x})$. As I claim all the time, the deep generative modeling paradigm is useful not only because we can synthesize nice-looking images but rather because an AI system can assess the uncertainty of the surrounding environment and share this information with us or other AI systems. Since calculating the marginal distribution in EBMs is troublesome, it is doubtful we can use these models in many applications. However, it is an extremely interesting research direction, and figuring out how to efficiently calculate the partition function and how to efficiently sample from the model is crucial for training powerful EBMs.

References

1. Yann LeCun, Sumit Chopra, Raia Hadsell, M Ranzato, and F Huang. A tutorial on energy-based learning. *Predicting structured data*, 1(0), 2006.
2. David H Ackley, Geoffrey E Hinton, and Terrence J Sejnowski. A learning algorithm for Boltzmann machines. *Cognitive science*, 9(1):147–169, 1985.
3. Paul Smolensky. Information processing in dynamical systems: Foundations of harmony theory. Technical report, Colorado Univ at Boulder Dept of Computer Science, 1986.
4. Edwin T Jaynes. *Probability theory: The logic of science*. Cambridge university press, 2003.
5. Will Grathwohl, Kuan-Chieh Wang, Joern-Henrik Jacobsen, David Duvenaud, Mohammad Norouzi, and Kevin Swersky. Your classifier is secretly an energy based model and you should treat it like one. In *International Conference on Learning Representations*, 2019.
6. Max Welling and Yee W Teh. Bayesian learning via stochastic gradient Langevin dynamics. In *Proceedings of the 28th international conference on machine learning (ICML-11)*, pages 681–688. Citeseer, 2011.
7. Geoffrey E Hinton, Terrence J Sejnowski, et al. Learning and relearning in Boltzmann machines. *Parallel distributed processing: Explorations in the microstructure of cognition*, 1(282-317):2, 1986.
8. Benjamin Marlin, Kevin Swersky, Bo Chen, and Nando Freitas. Inductive principles for restricted Boltzmann machine learning. In *Proceedings of the thirteenth International Conference on Artificial Intelligence and Statistics*, pages 509–516, 2010.
9. Geoffrey E Hinton. Training products of experts by minimizing contrastive divergence. *Neural computation*, 14(8):1771–1800, 2002.
10. Tijmen Tieleman. Training restricted Boltzmann machines using approximations to the likelihood gradient. In *ICML*, pages 1064–1071, 2008.
11. Jakub M Tomczak and Adam Gonczarek. Learning invariant features using subspace restricted Boltzmann machine. *Neural Processing Letters*, 45(1):173–182, 2017.
12. Roland Memisevic and Geoffrey E Hinton. Learning to represent spatial transformations with factored higher-order Boltzmann machines. *Neural computation*, 22(6):1473–1492, 2010.

13. Aaron Courville, James Bergstra, and Yoshua Bengio. A spike and slab restricted Boltzmann machine. In *Proceedings of the fourteenth international conference on artificial intelligence and statistics*, pages 233–241. JMLR Workshop and Conference Proceedings, 2011.
14. Ruslan Salakhutdinov, Andriy Mnih, and Geoffrey Hinton. Restricted Boltzmann machines for collaborative filtering. In *Proceedings of the 24th international conference on Machine learning*, pages 791–798, 2007.
15. Kyung Hyun Cho, Tapani Raiko, and Alexander Ilin. Gaussian-Bernoulli deep Boltzmann machine. In *The 2013 International Joint Conference on Neural Networks (IJCNN)*, pages 1–7. IEEE, 2013.
16. Ilya Sutskever, Geoffrey E Hinton, and Graham W Taylor. The recurrent temporal restricted Boltzmann machine. In *Advances in Neural Information Processing Systems*, pages 1601–1608, 2009.
17. Graham W Taylor, Leonid Sigal, David J Fleet, and Geoffrey E Hinton. Dynamical binary latent variable models for 3d human pose tracking. In *2010 IEEE Computer Society Conference on Computer Vision and Pattern Recognition*, pages 631–638. IEEE, 2010.
18. Hugo Larochelle and Yoshua Bengio. Classification using discriminative restricted Boltzmann machines. In *Proceedings of the 25th international conference on Machine learning*, pages 536–543, 2008.
19. Hugo Larochelle, Michael Mandel, Razvan Pascanu, and Yoshua Bengio. Learning algorithms for the classification restricted Boltzmann machine. *The Journal of Machine Learning Research*, 13(1):643–669, 2012.
20. Jakub M Tomczak. Learning informative features from restricted Boltzmann machines. *Neural Processing Letters*, 44(3):735–750, 2016.
21. Jakub M Tomczak. On some properties of the low-dimensional Gumbel perturbations in the Perturb-and-MAP model. *Statistics & Probability Letters*, 115:8–15, 2016.
22. Jakub M Tomczak, Szymon Zaręba, Siamak Ravanbakhsh, and Russell Greiner. Low-dimensional perturb-and-map approach for learning restricted Boltzmann machines. *Neural Processing Letters*, 50(2):1401–1419, 2019.
23. Jascha Sohl-Dickstein, Peter B Battaglino, and Michael R DeWeese. New method for parameter estimation in probabilistic models: Minimum Probability Flow. *Physical review letters*, 107(22):220601, 2011.
24. Yang Song and Diederik P Kingma. How to train your energy-based models. *arXiv preprint arXiv:2101.03288*, 2021.
25. Yoshua Bengio. *Learning deep architectures for AI*. Now Publishers Inc, 2009.
26. Honglak Lee, Roger Grosse, Rajesh Ranganath, and Andrew Y Ng. Convolutional deep belief networks for scalable unsupervised learning of hierarchical representations. In *Proceedings of the 26th annual international conference on machine learning*, pages 609–616, 2009.
27. Ruslan Salakhutdinov. Learning deep generative models. *Annual Review of Statistics and Its Application*, 2:361–385, 2015.
28. Ruslan Salakhutdinov and Iain Murray. On the quantitative analysis of deep belief networks. In *Proceedings of the 25th international conference on Machine learning*, pages 872–879, 2008.
29. Geoffrey E Hinton and Ruslan R Salakhutdinov. Reducing the dimensionality of data with neural networks. *science*, 313(5786):504–507, 2006.
30. Max Welling and Yee Whye Teh. Approximate inference in Boltzmann machines. *Artificial Intelligence*, 143(1):19–50, 2003.
31. Jonathan S Yedidia, William T Freeman, and Yair Weiss. Constructing free-energy approximations and generalized belief propagation algorithms. *IEEE Transactions on information theory*, 51(7):2282–2312, 2005.
32. Martin J Wainwright, Tommi S Jaakkola, and Alan S Willsky. A new class of upper bounds on the log partition function. *IEEE Transactions on Information Theory*, 51(7):2313–2335, 2005.
33. Tamir Hazan and Tommi Jaakkola. On the partition function and random maximum a-posteriori perturbations. In *Proceedings of the 29th International Conference on International Conference on Machine Learning*, pages 1667–1674, 2012.

Chapter 7
Generative Adversarial Networks

7.1 Introduction

Once we discussed latent variable models, we claimed that they naturally define a generative process by first sampling latents $\mathbf{z} \sim p(\mathbf{z})$ and then generating observables $\mathbf{x} \sim p_\theta(\mathbf{x}|\mathbf{z})$. That is nice! However, the problem appears when we start thinking about training. To be more precise, the training objective is an issue. Why? Well, the probability theory tells us to *get rid of* all unobserved random variables by marginalizing them out. In the case of latent variable models, this is equivalent to calculating the (marginal) log-likelihood function in the following form:

$$\log p_\theta(\mathbf{x}) = \log \int p_\theta(\mathbf{x}|\mathbf{z})\ p(\mathbf{z})\ \mathrm{d}\mathbf{z}. \tag{7.1}$$

As we mentioned already in the section about VAEs (see Sect. 4.3), the problematic part is calculating the integral because it is not analytically tractable unless all distributions are Gaussian and the dependency between $\mathbf{x}$ and $\mathbf{z}$ is linear. However, let us forget for a moment about all these issues and take a look at what we can do here. First, we can approximate the integral using Monte Carlo samples from the prior $p(\mathbf{z})$ that yields

$$\log p_\theta(\mathbf{x}) = \log \int p_\theta(\mathbf{x}|\mathbf{z})\ p(\mathbf{z})\ \mathrm{d}\mathbf{z} \tag{7.2}$$

$$\approx \log \frac{1}{S} \sum_{s=1}^{S} p_\theta(\mathbf{x}|\mathbf{z}_s) \tag{7.3}$$

$$= \log \sum_{s=1}^{S} \exp\left(\log p_\theta(\mathbf{x}|\mathbf{z}_s)\right) - \log S \tag{7.4}$$

© The Author(s), under exclusive license to Springer Nature Switzerland AG 2022
J. M. Tomczak, *Deep Generative Modeling*,
https://doi.org/10.1007/978-3-030-93158-2_7

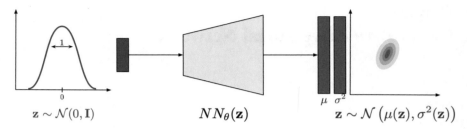

Fig. 7.1 A schematic representation of a density net

$$= \text{LogSumExp}_s \{ p_\theta(\mathbf{x}|\mathbf{z}_s) \} - \log S, \qquad (7.5)$$

where $\text{LogSumExp}_s \{ f(s) \} = \log \sum_{s=1}^{S} \exp(f(s))$ is the log-sum-exp function.

Assuming for a second that this is a good (i.e., a tight) approximation, we turn the problem of calculating the integral into a problem of sampling from the prior. For simplicity, we can assume a prior that is relatively easy to be sampled from, e.g., the standard Gaussian, $p(\mathbf{z}) = \mathcal{N}(\mathbf{z}|0, \mathbf{I})$. In other words, we need to model $p_\theta(\mathbf{x}|\mathbf{z})$ only, i.e., pick a parameterization for it. Guess what, we will use a neural network again! If we model images, then we can use the categorical distribution for the conditional likelihood and then a neural network parameterizes the probabilities. Or, if we use a Gaussian distribution like in the case of energy-based models or diffusion-based deep generative models, then $p_\theta(\mathbf{x}|\mathbf{z})$ could be Gaussian as well and the neural network outputs the variance and/or the mean. Since the log-sum-exp function is differentiable (and the application of the log-sum-exp trick makes it even numerically stable), there is no problem learning this model end-to-end! This approach is a precursor of many deep generative models and was dubbed *density networks* [1]; see Fig. 7.1 for a schematic representation of density networks.

Density networks are important and, unfortunately, underappreciated deep generative models. It is worth knowing them at least for three reasons. First, understanding how they work helps a lot to comprehend other latent variable models and how to improve them. Second, they serve as a great starting point for understanding the difference between *prescribed models* and *implicit models*. Third, they allow us to formulate a non-linear latent variable model and train it using backpropagation (or a gradient descent, in general).

Alright, so now you may have some questions because we made a few assumptions on the way that might have been pretty confusing. The main assumptions made here are the following:

- We need to specify the prior distribution $p(\mathbf{z})$, e.g., the standard Gaussian.
- We need to specify the form of the conditional likelihood $p(\mathbf{x}|\mathbf{z})$. Typically, people use the Gaussian distribution or a mixture of Gaussians. Hence, density nets are the *prescribed* models because we need to analytically formulate all distributions in advance.

As a result, we get the following:

- The objective function is the (approximated) log-likelihood function.
- We can optimize the objective using gradient-based optimization methods and the autograd tools.
- We can parameterize the conditional likelihood using deep neural networks.

However, we pay a great price for all the goodies coming from the formulation of the density networks:

- There is no analytical solution (except the case equivalent to the probabilistic PCA).
- We get an approximation of the log-likelihood function.
- We need a lot of samples from the prior to get a reliable approximation of the log-likelihood function.
- It suffers from the curse of dimensionality.

As you can see, the issue with dimensionality is especially limiting. What can we do with a model if it cannot be efficient for higher-dimensional problems? All interesting applications like image or audio analysis/synthesis are gone! So what can we do then? One possible direction is to stick to the prescribed models and apply variational inference (see Sect. 4.3). However, the other direction is to abandon the likelihood-based approach. I know, it sounds ridiculous, but it is possible and, *unfortunately*, works pretty well in practice.

7.2 Implicit Modeling with Generative Adversarial Networks (GANs)

Getting Rid of Kullback–Leibler

Let us think again what density networks tell us. First of all, they define a nice generative process: First sample latents and then generate observables. Clear! Then, for training, they use the (marginal) log-likelihood function. In other words, the log-likelihood function assesses the difference between a training datum and a generated object. To be even more precise, we first pick the specific probability distribution for the conditional likelihood $p_\theta(\mathbf{x}|\mathbf{z})$ that defines how to calculate the difference between the training point and the generated observables.

One may ask here whether there is a different fashion of calculating the *difference* between real data and generated objects. If we recall our considerations about hierarchical VAEs (see Sect. 4.5.2), learning of the likelihood-based models is equivalent to optimizing the Kullback-Leibler (KL) divergence between the empirical distribution and the model, $KL[p_{data}(\mathbf{x})||p_\theta(\mathbf{x})]$. The KL-based approach requires a well-behaved distribution because of the logarithms. Moreover, we can think of it as a *local* way of comparing the empirical distribution (i.e., given data) and the generated data (i.e., data generated by our prescribed model). By *local*, we mean considering one point at a time and then summing all individual errors instead of comparing samples (i.e., collections of individuals) that we can refer to as a *global* comparison. However, we do not need to stick to the KL divergence! Instead, we can use other metrics that look at a set of points (i.e., distributions represented by a set of points) like integral probability metrics [2] (e.g., the Maximum Mean Discrepancy [MMD] [3]) or use other divergences [4].

Still, all of the mentioned metrics rely on defining explicitly how we measure the error. The question is whether we can parameterize our loss function and learn it alongside our model. Since we talk all the time about neural networks, can we go even further and utilize a neural network to calculate differences?

Getting Rid of Prescribed Distributions
Alright, we agreed on the fact that the KL divergence is only one of many possible loss functions. Moreover, we asked ourselves whether we can use a learnable loss function. However, there is also one question floating in the air, namely, do we need to use the prescribed models in the first place? The reasoning is the following. Since we know that density networks take noise and turn them into distribution in the observable space, do we really need to output a full distribution? What if we return a single point? In other words, what if we define the conditional likelihood as Dirac's delta:

$$p_\theta(\mathbf{x}|\mathbf{z}) = \delta(\mathbf{x} - NN_\theta(\mathbf{z})). \qquad (7.6)$$

This is equivalent to saying that instead of a Gaussian (i.e., a mean and a variance), $NN_\theta(\mathbf{z})$ outputs the mean only. Interestingly, if we consider the marginal distribution over $\mathbf{x}$'s, we get nicely behaved distribution. To see that, let us first calculate the marginal distribution:

$$p_\theta(\mathbf{x}) = \int \delta(\mathbf{x} - NN_\theta(\mathbf{z})) \, p(\mathbf{z}) \, d\mathbf{z}. \qquad (7.7)$$

Then, let us understand what is going on! The marginal distribution is an infinite mixture of delta peaks. In other words, we take a single $\mathbf{z}$ and plot a peak (or a point in 2D, it is easier to imagine) in the observable space. We proceed to infinity and once we do that, the observable space will be covered by more and more points and some regions will be *denser* than the others. This kind of modeling a distribution is also known as *implicit modeling*.

So where is the problem then? Well, the problem in the prescribed modeling setting is that the term $\log \delta(\mathbf{x} - NN_\theta(\mathbf{z}))$ is ill-defined and cannot be used in many probability measures, including the KL-term, because we cannot calculate the loss function. Therefore, we can ask ourselves whether we can define our own loss function, perhaps? And, even more, parameterize it with neural networks! You must admit it sounds appealing! So how to accomplish that?

Adversarial Loss

Let us start with the following story. There is a con artist (a fraud) and a friend of the fraud (an expert) who knows a little about art. Moreover, there is a real artist who has passed away (e.g., Pablo Picasso). The fraud tries to mimic the style of Pablo Picasso as well as possible. The friend expert browses for paintings of Picasso and compares them to the paintings provided by the fraud. Hence, the fraud tries to fool his friend, while the expert tries to distinguish real paintings of Picasso from fakes. Over time, the fraud becomes better and better and the expert also learns how to decide whether a given painting is a fake. Eventually, and unfortunately to the world of art, work of the fraud may become indistinguishable from Picasso and the expert may be completely uncertain about the paintings and whether they are fakes.

Now, let us formalize this wicked game. We call the expert a *discriminator* that takes an object $\mathbf{x}$ and returns a probability whether it is *real* (i.e., coming from the empirical distribution), $D_\alpha : \mathcal{X} \to [0, 1]$. We refer to the fraud as a *generator* that takes noise and turns it into an object $\mathbf{x}$, $G_\beta : \mathcal{Z} \to \mathcal{X}$. All $\mathbf{x}$'s coming from the empirical distribution $p_{data}(\mathbf{x})$ are called *real* and all $\mathbf{x}$'s generated by $G_\beta(\mathbf{z})$ are dubbed *fake*. Then, we construct the objective function as follows:

- We have two sources of data: $\mathbf{x} \sim p_\theta(\mathbf{x}) = \int G_\beta(\mathbf{z}) \, p(\mathbf{z}) \, d\mathbf{z}$ and $\mathbf{x} \sim p_{data}(\mathbf{x})$.
- The discriminator solves the classification task by assigning 0 to all fake datapoints and 1 to all real datapoints.
- Since the discriminator can be seen as a classifier, we can use the binary cross-entropy loss function in the following form:

$$\ell(\alpha, \beta) = \mathbb{E}_{\mathbf{x} \sim p_{real}} \left[\log D_\alpha(\mathbf{x}) \right] + \mathbb{E}_{\mathbf{z} \sim p(\mathbf{z})} \left[\log \left(1 - D_\alpha \left(G_\beta(\mathbf{z}) \right) \right) \right]. \qquad (7.8)$$

The left part corresponds to the real data source, and the right part contains the fake data source.

- We try to maximize $\ell(\alpha, \beta)$ with respect to α (i.e., the discriminator). In plain words, we want the discriminator to be as good as possible.
- The generator tries to fool the discriminator and, thus, it tries to minimize $\ell(\alpha, \beta)$ with respect to β (i.e., the generator).

Eventually, we face the following learning objective:

$$\min_\beta \max_\alpha \mathbb{E}_{\mathbf{x} \sim p_{real}} \left[\log D_\alpha(\mathbf{x}) \right] + \mathbb{E}_{\mathbf{z} \sim p(\mathbf{z})} \left[\log \left(1 - D_\alpha \left(G_\beta(\mathbf{z}) \right) \right) \right]. \qquad (7.9)$$

We refer to $\ell(\alpha, \beta)$ as the *adversarial loss* since there are two actors trying to achieve two opposite goals.

GANs

Let us put everything together:

- We have a generator that turns noise into fake data.
- We have a discriminator that classifies given input as either fake or real.
- We parameterize the generator and the discriminator using deep neural networks.

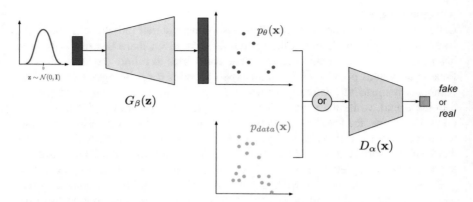

Fig. 7.2 A schematic representation of GANs. Please note the part of the generator and its resemblance to density networks

- We learn the neural networks using the adversarial loss (i.e., we optimize the min–max problem).

The resulting class of models is called Generative Adversarial Networks (GANs) [5]. In Fig. 7.2, we present the idea of GANs and how they are connected to density networks. Notice that the generator part constitutes an implicit distribution, i.e., a distribution from an unknown family of distributions, and its analytical form is unknown as well; however, we can sample from it.

7.3   Implementing GANs

Believe me or not, but we have all components to implement GANs. Let us look into all of them step-by-step. In fact, the easiest way to understand them is to implement them.

Generator
The first part is the *generator*, $G_\beta(\mathbf{z})$, which is simply a deep neural network. The code for a class of the generator is presented below. Notice that we distinguish between a function for generating, namely, transforming $\mathbf{z}$ to $\mathbf{x}$, and sampling that first samples $\mathbf{z} \sim \mathcal{N}(0, \mathbf{I})$ and then calls `generate`.

```
class Generator(nn.Module):
    def __init__(self, generator_net, z_size):
        super(Generator, self).__init__()

        # We need to init the generator neural net.
        self.generator_net = generator_net
        # We also need to know the size of the latents.
```

```
8       self.z_size = z_size
9
10   def generate(self, z):
11       # Generating for given z is equivalent to applying the
     neural net.
12       return self.generator_net(z)
13
14   def sample(self, batch_size=16):
15       # For sampling, we need to sample first latents.
16       z = torch.randn(batch_size, self.z_size)
17       return self.generate(z)
18
19   def forward(self, z=None):
20       if z is None:
21           return self.sample()
22       else:
23           return self.generate(z)
```

Listing 7.1   A Generator class

Discriminator

The second component is the *discriminator*. Here, the code is even simpler because it consists of a single neural network. The code for a class of the discriminator is provided below:

```
1  class Discriminator(nn.Module):
2      def __init__(self, discriminator_net):
3          super(Discriminator, self).__init__()
4          # We need to init the discriminator neural net.
5          self.discriminator_net = discriminator_net
6
7      def forward(self, x):
8          # The forward pass is just about applying the neural net.
9          return self.discriminator_net(x)
```

Listing 7.2   A Discriminator class

GAN

Now, we are ready to combine these two components. In our implementation, a GAN outputs the adversarial loss either for the generator or for the discriminator. Maybe the code below is overkill; however, it is better to write a few more lines and properly understand what is going on than applying some unclear tricks.

```
1  class GAN(nn.Module):
2      def __init__(self, generator, discriminator, EPS=1.e-5):
3          super(GAN, self).__init__()
4
5          print('GAN by JT.')
6
7          # To put everything together, we need the generator and
8          # the discriminator. NOTE: Both are instances of classes!
9          self.generator = generator
```

```
10        self.discriminator = discriminator
11
12        # For numerical issue, we introduce a small epsilon.
13        self.EPS = EPS
14
15    def forward(self, x_real, reduction='avg', mode='
      discriminator'):
16        # The forward pass calculates the adversarial loss.
17        # More specifically, either its part for the generator or
18        #  the part for the discriminator.
19        if mode == 'generator':
20            # For the generator, we first sample FAKE data.
21            x_fake_gen = self.generator.sample(x_real.shape[0])
22
23            # Then, we calculate outputs of the discriminator for
      the FAKE data.
24            # NOTE: We clamp here for the numerical stability
      later on.
25            d_fake = torch.clamp(self.discriminator(x_fake_gen),
      self.EPS, 1. - self.EPS)
26
27            # The loss for the generator is log(1 - D(G(z))).
28            loss = torch.log(1. - d_fake)
29
30        elif mode == 'discriminator':
31            # For the discriminator, we first sample FAKE data.
32            x_fake_gen = self.generator.sample(x_real.shape[0])
33
34            # Then, we calculate outputs of the discriminator for
      the FAKE data.
35            # NOTE: We clamp for the numerical stability later on
      .
36            d_fake = torch.clamp(self.discriminator(x_fake_gen),
      self.EPS, 1. - self.EPS)
37
38            # Moreover, we calculate outputs of the discriminator
      for the REAL data.
39            # NOTE: We clamp for... the numerical stability (
      again).
40            d_real = torch.clamp(self.discriminator(x_real), self
      .EPS, 1. - self.EPS)
41
42            # The final loss for the discriminator is log(1 - D(G
      (z))) + log D(x).
43            # NOTE: We take the minus sign because we MAXIMIZE
      the adversarial loss wrt
44            # discriminator, so we MINIMIZE the negative
      adversarial loss wrt discriminator.
45            loss = -(torch.log(d_real) + torch.log(1. - d_fake))
46
47        if reduction == 'sum':
48            return loss.sum()
49        else:
50            return loss.mean()
```

```
51
52    def sample(self, batch_size=64):
53        return self.generator.sample(batch_size=batch_size)
```

Listing 7.3 A GAN class

Examples of architectures for a generator and a discriminator are presented in the code below:

```
1  # First, we initialize the generator and the discriminator
2  # —generator
3  generator_net = nn.Sequential(nn.Linear(L, M), nn.ReLU(),
4                                nn.Linear(M, D), nn.Tanh())
5
6  generator = Generator(generator_net, z_size=L)
7
8  # —discriminator
9  discriminator_net = nn.Sequential(nn.Linear(D, M), nn.ReLU(),
10                                   nn.Linear(M, 1), nn.Sigmoid())
11
12 discriminator = Discriminator(discriminator_net)
13
14 # Eventually, we initialize the full model
15 model = GAN(generator=generator, discriminator=discriminator)
```

Listing 7.4 Examples of architectures

Training
One might think that the training procedure for GANs is more complicated than for any of the likelihood-based models. However, it is not the case. The only difference is that we need **two optimizers** instead of one. An example of a code with a training loop is presented below:

```
1  # We use two optimizers:
2  # optimizer_dis — an optimizer that takes the parameters of the
       discriminator
3  # optimizer_gen — an optimizer that takes the parameters of the
       generator
4  for indx_batch, batch in enumerate(training_loader):
5
6      # —Discriminator
7      # Notice that we call our model with the 'discriminator' mode
       .
8      loss_dis = model.forward(batch, mode='discriminator')
9
10     optimizer_dis.zero_grad()
11     optimizer_gen.zero_grad()
12     loss_dis.backward(retain_graph=True)
13     optimizer_dis.step()
14
15     # —Generator
16     # Notice that we call our model with the 'generator' mode.
17     loss_gen = model.forward(batch, mode='generator')
```

```
18
19      optimizer_dis.zero_grad()
20      optimizer_gen.zero_grad()
21      loss_gen.backward(retain_graph=True)
22      optimizer_gen.step()
```

Listing 7.5   A training loop

Results and Comments
In the experiments, we normalized images and scaled them to $[-1, 1]$ as we did for
EBMs. The full code (with auxiliary functions) that you can play with is available
here: https://github.com/jmtomczak/intro_dgm. After running it, you can expect
similar results to those in Fig. 7.3.

In the previous chapters, we did not comment on the results. However, we make
an exception here. Please note, my curious reader, that now we do not have a nicely
converging objective. On the contrary, the adversarial loss or its generating part
is jumping all over the place. That is a known fact following from the min–max
optimization problem. Moreover, the loss is learnable now so it is troublesome to
say where the optimal solution is since we update the loss function as well.

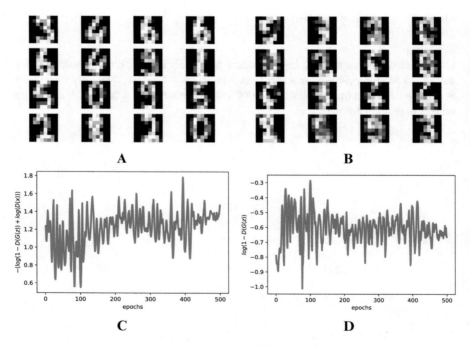

Fig. 7.3   Examples of results after running the code for GANs. (**a**) Real images. (**b**) Fake images.
(**c**) The validation curve for the discriminator. (**d**) The validation curve for the generator

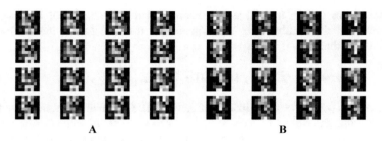

A   B

Fig. 7.4 Generated images after (**a**) 10 epochs of training and (**b**) 50 epochs of training

Another important piece of information is that training GANs is indeed a pain. First, it is hard to decipher and properly understand the values of the adversarial loss. Second, learning is rather slow and requires many iterations (by many I mean hundreds if not thousands). If you look into generations in the first few epochs (e.g., see Fig. 7.4), you may be discouraged because a model may seem to overfit. That is the problem, we must be really patient to see whether we are on the good track. Moreover, you may also need to pay special attention to hyperparameters, e.g., learning rates. It requires a bit of experience or simply time to play around with learning rate values in your problem.

Once you get through learning GANs, the reward is truly amazing. In the presented problem, with extremely simple neural nets, we are able to synthesize digits of high quality. That's the biggest advantage of GANs!

7.4   There Are Many GANs Out There!

Since the publication of the seminal paper on GANs [5] (however, the idea of the adversarial problem could be traced back to [6]), there was a flood of GAN-based ideas and papers. I would not even dare to mention a small fraction of them. The field of implicit modeling with GANs is growing constantly. I will try to point to a few important papers:

- *Conditional GANs*: An important extension of GANs is allowing them to generate data conditionally [7].
- *GANs with encoders*: An interesting question is whether we can extend conditional GANs to a framework with encoders. It turns out that it is possible; see BiGAN [8] and ALI [9] for details.
- *StyleGAN* and *CycleGAN*: The flexibility of GANs could be utilized in formulating specialized image synthesizers. For instance, StyleGAN is formulated in such a way to transfer style between images [10], while CycleGAN tries to "translate" one image into another, e.g., a horse into a zebra [11].

- *Wasserstein GANs*: In [12] it was claimed that the adversarial loss could be formulated differently using the Wasserstein distance (a.k.a. the earth-mover distance), that is:

$$\ell_W(\alpha, \beta) = \mathbb{E}_{\mathbf{x} \sim p_{real}} [D_\alpha(\mathbf{x})] - \mathbb{E}_{\mathbf{z} \sim p(\mathbf{z})} [D_\alpha (G_\beta(\mathbf{z}))]. \tag{7.10}$$

  where $D_\alpha(\cdot)$ must be a 1-Lipschitz function. The simpler way to achieve that is by clipping the weight of the discriminator to some small value c. Alternatively, *spectral normalization* could be applied [13] by using the power iteration method. Overall, constraining the discriminator to be a 1-Lipshitz function stabilizes training; however, it is still hard to comprehend the learning process.
- *f-GANs*: The Wasserstein GAN indicated that we can look elsewhere for alternative formulations of the adversarial loss. In [14], it is advocated to use f-divergences for that.
- *Generative Moment Matching Networks* [15, 16]: As mentioned earlier, we could use other metrics instead of the likelihood function. We can fix the discriminator and define it as the Maximum Mean Discrepancy with a given kernel function. The resulting problem is simpler because we do not train the discriminator and, thus, we get rid of the cumbersome min–max optimization. However, the final quality of synthesized images is typically poorer.
- *Density difference vs. Density ratio*: An interesting perspective is presented in [17, 18] where we can see various GANs either as a difference of densities or as a ratio of densities. I refer to the original papers for further details.
- *Hierarchical implicit models*: The idea of defining implicit models could be extended to hierarchical models [19].
- *GANs and EBMs*: If you recall the EBMs, you may notice that there is a clear connection between the adversarial loss and the logarithm of the Boltzmann distribution. In [20, 21] it was noticed that introducing a variational distribution over observables, $q(\mathbf{x})$, leads to the following objective:

$$\mathcal{J}(\mathbf{x}) = \mathbb{E}_{\mathbf{x} \sim p_{data}(\mathbf{x})} [E(\mathbf{x})] - \mathbb{E}_{\mathbf{x} \sim q(\mathbf{x})} [E(\mathbf{x})] + \mathbb{H}[q(\mathbf{x})], \tag{7.11}$$

  where $E(\cdot)$ is the energy function and $\mathbb{H}[\cdot]$ is the entropy. The problem again boils down to the min–max optimization problem, namely, minimizing with respect to the energy function and maximizing with respect to the variational distribution. The second difference between the adversarial loss and the variational lower bound here is the entropy term that is typically intractable.
- *What GAN to use?*: That is the question! Interestingly, it seems that training GANs greatly depends on the initialization and the neural nets rather than the adversarial loss or other tricks. You can read more about it in [22].
- *Training instabilities*: The main problem of GANs is unstable learning and a phenomenon called *mode collapse*, namely, a GAN samples beautiful images but only from some regions of the observable space. This problem has been studied for a long time by many (e.g., [23–25]); however, it still remains an open question.

- *Prescribed GANs*: Interestingly, it is possible to still calculate the likelihood for a GAN! See [26] for more details.
- *Regularized GANs*: There are many ideas to regularize GANs to achieve specific goals. For instance, InfoGAN aims for learning disentangled representations by introducing the mutual-information-based regularizer [27].

Each of these ideas constitutes a separate research direction followed by thousands of researchers. If you are interested in pursuing any of these, I suggest picking one of the paper mentioned here and start digging!

References

1. David JC MacKay and Mark N Gibbs. Density networks. *Statistics and neural networks: advances at the interface*, pages 129–145, 1999.
2. Bharath K Sriperumbudur, Kenji Fukumizu, Arthur Gretton, Bernhard Schölkopf, and Gert RG Lanckriet. On integral probability metrics, ϕ-divergences and binary classification. *arXiv preprint arXiv:0901.2698*, 2009.
3. Arthur Gretton, Karsten Borgwardt, Malte Rasch, Bernhard Schölkopf, and Alex Smola. A kernel method for the two-sample-problem. *Advances in Neural Information Processing Systems*, 19:513–520, 2006.
4. Tim Van Erven and Peter Harremos. Rényi divergence and Kullback-Leibler divergence. *IEEE Transactions on Information Theory*, 60(7):3797–3820, 2014.
5. Ian J Goodfellow, Jean Pouget-Abadie, Mehdi Mirza, Bing Xu, David Warde-Farley, Sherjil Ozair, Aaron Courville, and Yoshua Bengio. Generative adversarial networks. *arXiv preprint arXiv:1406.2661*, 2014.
6. Jürgen Schmidhuber. Making the world differentiable: On using fully recurrent self-supervised neural networks for dynamic reinforcement learning and planning in non-stationary environments. *Institut für Informatik, Technische Universität München. Technical Report FKI-126*, 90, 1990.
7. Mehdi Mirza and Simon Osindero. Conditional generative adversarial nets. *arXiv preprint arXiv:1411.1784*, 2014.
8. Jeff Donahue, Philipp Krähenbühl, and Trevor Darrell. Adversarial feature learning. *arXiv preprint arXiv:1605.09782*, 2016.
9. Vincent Dumoulin, Ishmael Belghazi, Ben Poole, Olivier Mastropietro, Alex Lamb, Martin Arjovsky, and Aaron Courville. Adversarially learned inference. *arXiv preprint arXiv:1606.00704*, 2016.
10. Tero Karras, Samuli Laine, and Timo Aila. A style-based generator architecture for generative adversarial networks. In *Proceedings of the IEEE/CVF Conference on Computer Vision and Pattern Recognition*, pages 4401–4410, 2019.
11. Jun-Yan Zhu, Taesung Park, Phillip Isola, and Alexei A Efros. Unpaired image-to-image translation using cycle-consistent adversarial networks. In *Proceedings of the IEEE international conference on computer vision*, pages 2223–2232, 2017.
12. Martin Arjovsky, Soumith Chintala, and Léon Bottou. Wasserstein generative adversarial networks. In *International conference on machine learning*, pages 214–223. PMLR, 2017.
13. Takeru Miyato, Toshiki Kataoka, Masanori Koyama, and Yuichi Yoshida. Spectral normalization for generative adversarial networks. *arXiv preprint arXiv:1802.05957*, 2018.
14. Sebastian Nowozin, Botond Cseke, and Ryota Tomioka. f-GAN: Training generative neural samplers using variational divergence minimization. In *Advances in Neural Information Processing Systems*, pages 271–279, 2016.

15. Gintare Karolina Dziugaite, Daniel M Roy, and Zoubin Ghahramani. Training generative neural networks via maximum mean discrepancy optimization. In *Proceedings of the Thirty-First Conference on Uncertainty in Artificial Intelligence*, pages 258–267, 2015.
16. Yujia Li, Kevin Swersky, and Rich Zemel. Generative moment matching networks. In *International Conference on Machine Learning*, pages 1718–1727. PMLR, 2015.
17. Ferenc Huszár. Variational inference using implicit distributions. *arXiv preprint arXiv:1702.08235*, 2017.
18. Shakir Mohamed and Balaji Lakshminarayanan. Learning in implicit generative models. *arXiv preprint arXiv:1610.03483*, 2016.
19. Dustin Tran, Rajesh Ranganath, and David M Blei. Hierarchical implicit models and likelihood-free variational inference. *Advances in Neural Information Processing Systems*, 2017:5524–5534, 2017.
20. Taesup Kim and Yoshua Bengio. Deep directed generative models with energy-based probability estimation. *arXiv preprint arXiv:1606.03439*, 2016.
21. Shuangfei Zhai, Yu Cheng, Rogerio Feris, and Zhongfei Zhang. Generative adversarial networks as variational training of energy based models. *arXiv preprint arXiv:1611.01799*, 2016.
22. Mario Lucic, Karol Kurach, Marcin Michalski, Sylvain Gelly, and Olivier Bousquet. Are GANs created equal? A large-scale study. *Advances in Neural Information Processing Systems*, 31, 2018.
23. Tim Salimans, Ian Goodfellow, Wojciech Zaremba, Vicki Cheung, Alec Radford, and Xi Chen. Improved techniques for training GANs. *Advances in neural information processing systems*, 29:2234–2242, 2016.
24. Lars Mescheder, Andreas Geiger, and Sebastian Nowozin. Which training methods for GANs do actually converge? In *International Conference on Machine Learning*, pages 3481–3490. PMLR, 2018.
25. Augustus Odena, Christopher Olah, and Jonathon Shlens. Conditional image synthesis with auxiliary classifier GANs. In *International conference on machine learning*, pages 2642–2651. PMLR, 2017.
26. Adji B Dieng, Francisco JR Ruiz, David M Blei, and Michalis K Titsias. Prescribed generative adversarial networks. *arXiv preprint arXiv:1910.04302*, 2019.
27. Xi Chen, Yan Duan, Rein Houthooft, John Schulman, Ilya Sutskever, and Pieter Abbeel. Info-GAN: Interpretable representation learning by information maximizing generative adversarial nets. In *Proceedings of the 30th International Conference on Neural Information Processing Systems*, pages 2180–2188, 2016.

Chapter 8
Deep Generative Modeling for Neural Compression

8.1 Introduction

In December 2020, Facebook reported having around 1.8 billion daily active users and around 2.8 billion monthly active users [1]. Assuming that users uploaded, on average, a single photo each day, the resulting volume of data would give a very rough (let me stress it: **a very rough**) estimate of around 3000TB of new images per day. This single case of Facebook alone already shows us potential great costs associated with storing and transmitting data. In the digital era we can simply say this: efficient and effective manner of handling data (i.e., **faster** and **smaller**) means more money in the pocket.

The most straightforward way of dealing with these issues (i.e., smaller and faster) is based on applying compression, and, in particular, image compression algorithms (codecs) that allow us to decrease the size of an image. Instead of changing infrastructure, we can efficiently and effectively store and transmit images by making them simply smaller! Let us be honest, the more we compress an image, the more and faster we can send and the less disk memory we need!

If we think of image compression, probably the first association is JPEG or PNG, standards used on the daily basis by everyone. I will not go into details of these standards (e.g., see [2, 3] for an introduction) but what it is important to know is that they use some pre-defined math like Discrete Cosine Transform. The main advantage of the standard codecs like JPEG is that they are interpretable, i.e., all steps are hand-designed and their behavior could be predicted. However, this comes at the cost of insufficient flexibility that could drastically decrease their performance. So how we can increase the flexibility of transformations? Any idea? Anyone? Do I hear deep learning [4, 5]? Indeed! Many of today's image compression algorithms are enhanced by neural networks.

The emerging field of compression algorithms using neural networks is called **neural compression**. Neural compression becomes a leading trend in developing new codecs where neural networks replace parts of the standard codecs [6], or

J. M. Tomczak, *Deep Generative Modeling*,
https://doi.org/10.1007/978-3-030-93158-2_8

neural-based codecs are trained [7] together with quantization [8] and entropy coding [9–12]. We will discuss the general compression scheme in detail in the next subsection but here it is important to understand why deep generative modeling is important in the context of neural compression. The answer was given a long time ago by Claude Shannon who showed in [13] that (informally):

> The length of a message representing a source data is proportional to the entropy of this data.

We do not know the entropy of data because we do not know the probability distribution of data, $p(\mathbf{x})$, but we can estimate it using one of the deep generative models we have discussed so far! Because of that, recently, there is an increasing interest in using deep generative modeling to improve neural compression. We can use deep generative models for modeling probability distribution for entropy coders [9–12], but also to significantly increase the final reconstruction and compression quality by incorporating new schemes for inference [14] and reconstruction [15].

8.2   General Compression Scheme

Before we jump into neural compression, it is beneficial (and educational, don't be afraid of refreshing the basics!) to remind ourselves what is image (or data, in general) compression. We can distinguish two image compression approaches [16], namely:

- *lossless compression*: a method that preserves all information and reconstructions are error-free,
- *lossy compression*: information is not preserved completely by a compression method.

A general recipe for designing a compression algorithm relies on devising a uniquely decodable code whose expected length is as close as possible to the entropy of the data [13]. The general compression system consists of two components [16, 17]: an **encoder** and a **decoder**. Please do not think of it as a deterministic VAE because it is not the same. There are some similarities but in the compression task, we are really interested in sending the **bitstream** while in VAEs we typically do not care at all. We play with floats and say about codes just to make it more intuitive but it requires a few extra steps to turn it into a "real" compression scheme. We are going to explain these extra in the next sections. Alright, let's start!

Encoder
In the encoder, the goal is to transform an image into a discrete signal. It is important to understand that the signal does not necessarily need to be binary. The transformation we use could be invertible, however, it is not a requirement. If the transformation is invertible, then we can use its inverse in the decoder and, in principle, we can talk about lossless compression. Why? Take a look at the flow-

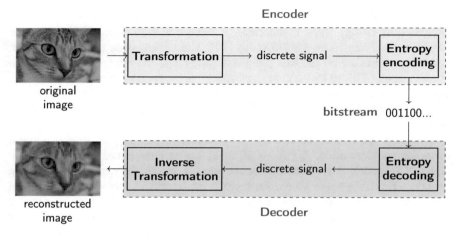

Fig. 8.1 A general image compression system (a codec)

based models where we discussed invertibility. However, if the transformation is not invertible, then some information is lost and we land in the group of lossy compression methods. Next, after applying the transformation to the input image, the discrete signal is encoded into a bitstream in a lossless manner. In other words, discrete symbols are mapped to binary variables (bits). Typically, entropy coders utilize information about the probability of symbol occurrence, e.g., Huffman coders or arithmetic coders [16]. This is important to understand that for many entropy coders we need to know $p(\mathbf{x})$ and here we can use deep generative models.

Decoder
Once the message (i.e., the bits) is sent and received, the bitstream is decoded to the discrete signal by the entropy decoder. The entropy decoder is the inversion of the entropy encoder. Entropy coding methods allow us to receive the original symbols from the bitstream. Eventually, an inverse transformation (not necessarily the inversion of the encoder transformation) is applied to reconstruct the original image.

The Full Scheme
Please see Fig. 8.1 for a general scheme of a compression system (a codec). The standard codecs mainly utilize multi-scale image decomposition like wavelet representation [3, 18] that are further quantized. Considering a specific discrete transformation (e.g., Discrete Cosine Transform (DCT)) results in a specific codec (e.g., JPEG [2]).

The Objective
The final performance of a codec is evaluated in terms of reconstruction error and compression ratio. The reconstruction error is called **distortion** measure and is calculated as a difference between the original image and the reconstructed image

using the mean square error (MSE) (typically, the peak signal-to-noise ratio $PSNR$ expressed as $10\log_{10}\frac{255^2}{MSE}$ is reported) or perceptual metrics like the multi-scale structure similarity index (MS-$SSIM$) [19]. The compression ratio, called **rate**, is usually expressed by the **bits per pixel (bpp)**, i.e., the total size in bits of the encoder output divided by the total size in pixels of the encoder input [17]. Typically, the performance of codecs is compared by inspecting the rate-distortion plane (i.e., plotting curves on a plane with the rate on the x-axis and the distortion on the y-axis).

Formally, we assume an auto-ecoder architecture (see Fig. 8.1 again) with an encoding transformation, $f_e : \mathcal{X} \rightarrow \mathcal{Y}$, that takes an input **x** and returns a discrete signal **y** (a code). After sending the message, a reconstruction $\hat{\mathbf{x}}$ is given by a decoder, $f_d : \mathcal{Y} \rightarrow \mathcal{X}$. Moreover, there is an (adaptive) entropy coding model that learns the distribution $p(\mathbf{y})$ and is further used to turn the discrete signal **y** into a bitstream by an entropy coder (e.g., Huffman coding, arithmetic coding). If a compression method has any adaptive (hyper)parameters, it could be learned by optimizing the following objective function:

$$\mathcal{L}(\mathbf{x}) = d\left(\mathbf{x}, \hat{\mathbf{x}}\right) + \beta r\left(\mathbf{y}\right), \tag{8.1}$$

where $d(\cdot, \cdot)$ is the **distortion** measure (e.g., PSNR, MS-SSIM), and $r(\cdot)$ is the **rate** measure (e.g., $r\left(\mathbf{y}\right) = -\ln p(\mathbf{y})$), $\beta > 0$ is a weighting factor controlling the balance between rate and distortion. Notice that distortion requires both the encoder and the decoder, and rate requires the encoder, and the entropy model.

8.3 A Short Detour: JPEG

We have discussed all necessary concepts of image compression and now we can delve into neural compression. However, before we do that, the first question we face is whether we can get any benefit from using neural networks for compression and where and how we can use them in this context. As mentioned already, standard codecs utilize a series of pre-defined transformations and mathematical operations. But how does it work?

Let us quickly discuss one of the most commonly used codecs: JPEG. In the JPEG codec, an RGB image is first linearly transformed to the YCbCr format:

$$\begin{bmatrix} Y \\ Cb \\ Cr \end{bmatrix} = \begin{bmatrix} 0 \\ 128 \\ 128 \end{bmatrix} + \begin{bmatrix} 0.299 & 0.587 & 0.114 \\ -0.168736 & -0.331264 & 0.5 \\ 0.5 & -0.48688 & -0.081312 \end{bmatrix} \begin{bmatrix} R \\ G \\ B \end{bmatrix} \tag{8.2}$$

Then, the Cb and Cr channels are downscaled, typically two or three times (the first compression stage). After that, each channel is split into, e.g., 8×8 blocks, and fed to the discrete cosine transform (DCT) that is eventually quantized (the

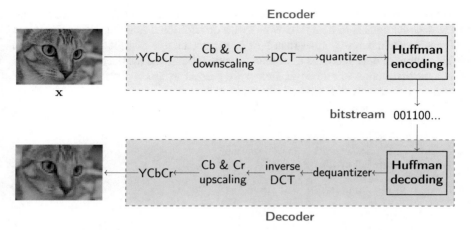

Fig. 8.2 A JPEG compression system

second compression stage). After all, the Huffman coding could be used. To decode the signal, the inverse DCT is used, the Cb and Cr channels are upscaled and the RGB representation is recovered. The whole system is presented in Fig. 8.2. As you can notice, each step is easy to follow and if you know how DCT works, the whole procedure is a white box. There are some hyperparameters but, again, they have a very clear interpretation (e.g., how many times the Cb and Cr channels are downscaled, the size of blocks).

8.4   Neural Compression: Components

Alright, we know now how a standard codec works. However, one of the problems with standard codecs is that they are not necessarily flexible. One may ask whether DCT is the optimal transformation for all images. The answer is, with high probability, no. If we are willing to give up the nicely designed white box, we can turn it into a black box by replacing all mathematical operations with neural networks. The potential gain is increased flexibility and potentially better performance (both in terms of distortion and rate).

We need to remember though that learning neural network requires the **differentiability** of the whole approach. However, we require a discrete output of the neural network that could break the backpropagation! For this purpose, we must formulate a **differentiable quantization procedure**. Additionally, to obtain a powerful model, we need an adaptive entropy coding model. This is an important component of the neural compression pipeline because it not only optimizes the compression ratio (i.e., the rate) but also helps to learn a useful codebook. Next, we will discuss both components in detail.

Encoders and Decoders   In neural compression, unlike in VAEs, the encoder and
the decoder consist of neural networks with no additional functions. As a result, we
focus on architectures rather than how to parameterize a distribution, for instance.
The output of the encoder is a continuous code (floats) and the output of the decoder
is a reconstruction of an image. Below we present PyTroch classes for an encoder
and a decoder with examples of neural networks.

```python
1  # The encoder is simply a neural network that takes an image and
       outputs a corresponding code.
2  class Encoder(nn.Module):
3      def __init__(self, encoder_net):
4          super(Encoder, self).__init__()
5
6          self.encoder = encoder_net
7
8      def encode(self, x):
9          h_e = self.encoder(x)
10         return h_e
11
12     def forward(self, x):
13         return self.encode(x)
14
15 # The decoder is simply a neural network that takes a quantized
       code and returns an image.
16 class Decoder(nn.Module):
17     def __init__(self, decoder_net):
18         super(Decoder, self).__init__()
19
20         self.decoder = decoder_net
21
22     def decode(self, z):
23         h_d = self.decoder(z)
24         return h_d
25
26     def forward(self, z, x=None):
27         x_rec = self.decode(z)
28         return x_rec
```

Listing 8.1 Classes for an encoder and a decoder

```python
1  # ENCODER
2  e_net = nn.Sequential(
3          nn.Linear(D, M*2), nn.BatchNorm1d(M*2), nn.ReLU(),
4          nn.Linear(M*2, M), nn.BatchNorm1d(M), nn.ReLU(),
5          nn.Linear(M, M//2), nn.BatchNorm1d(M//2), nn.ReLU(),
6          nn.Linear(M//2, C))
7
8  encoder = Encoder(encoder_net=e_net)
9
10 # DECODER
11 d_net = nn.Sequential(
12         nn.Linear(C, M//2), nn.BatchNorm1d(M//2), nn.ReLU(),
13         nn.Linear(M//2, M), nn.BatchNorm1d(M), nn.ReLU(),
```

```
14            nn.Linear(M, M*2), nn.BatchNorm1d(M*2), nn.ReLU(),
15            nn.Linear(M*2, D))
16
17 decoder = Decoder(decoder_net=d_net)
```

Listing 8.2 Examples of neural networks for the encoder and the decoder

Differentiable Quantization The problem with utilizing neural networks in the compression context is that we must ensure training by backpropagation that is equivalent to using only differentiable operations. Unfortunately, working with discrete outputs of a neural network breaks differentiability and requires applying approximations of gradients (e.g., the Straight-Through Estimator). However, we can use quantization of codes $\mathbf{y}$ and make it differentiable with relatively simple tricks.

We assume that the encoder gives us a code $\mathbf{y} \in \mathbb{R}^M$. Moreover, we assume that there is a codebook $\mathbf{c} \in \mathbb{R}^K$. We can think of the codebook as a vector of additional parameters (yes, parameters, we can also learn it!). Now, the whole idea relies on quantizing $\mathbf{y}$ to values in the codebook $\mathbf{c}$. Easy right? Well, it is easy but it still tells us nothing. Quantizing in this context means that we will take every element in $\mathbf{y}$ and find the closest value in the codebook and replace it with this codebook value. We can implement it using matrix calculus in the following manner. First, we repeat $\mathbf{y}$ K-times and we repeat $\mathbf{c}$ M-times that gives us two matrices: $\mathbf{Y} \in \mathbb{R}^{M \times K}$ and $\mathbf{C} \in \mathbb{R}^{M \times K}$. Now, we can calculate a similarity matrix, for instance: $\mathbf{S} = \exp\{-\sqrt{(\mathbf{Y} - \mathbf{C})^2}\} \in \mathbb{R}^{M \times K}$. The matrix $\mathbf{S}$ has the highest value where the m-th value of $\mathbf{y}$, $\mathbf{y}_m$, is closest to the k-th vale of $\mathbf{c}$, $\mathbf{c}_k$. So far so good, all operations are differentiable. However, there is no quantization here (i.e., values are not discrete). Since we have the similarity matrix $\mathbf{S} \in \mathbb{R}^{M \times K}$ and we can apply the softmax non-linearity with temperature to the second dimension of $\mathbf{S}$, namely $\hat{\mathbf{S}} = \text{softmax}_2(\tau \cdot \mathbf{S})$ (here, the subscript denotes that we calculate the softmax w.r.t the second dimension) where $\tau \gg 1$ (e.g., $\tau = 10^7$). Since we apply the softmax to the similarity matrix multiplied by a very large number, the resulting matrix, $\hat{\mathbf{S}}$, will still consist of floats but numerically these values will be 0's and a single 1. An example of this kind of quantization is presented in Fig. 8.3.

Importantly, the softmax non-linearity is differentiable and, eventually, the whole procedure is differentiable. In the end, we can calculate quantized codes by multiplying the codebook with the 0-1 similarity matrix, namely

$$\hat{\mathbf{y}} = \hat{\mathbf{S}}\mathbf{c}. \tag{8.3}$$

The resulting code, $\hat{\mathbf{y}}$, consists of values from the codebook only.

We can ask ourselves whether indeed we gain anything because values of $\hat{\mathbf{y}}$ are still floats. So, in other words, where is the discrete signal we want to turn into the bitstream? We can answer it in two ways. First, there are only K possible values in the codebook. Hence, the values are discrete but represented by a finite number of floats. Second, the real magic happens when we calculate the matrix $\hat{\mathbf{S}}$. Notice that this matrix is indeed discrete! In each row, there is a single position with 1

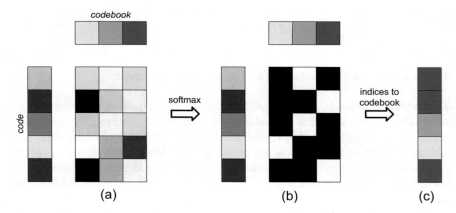

Fig. 8.3 An example of the quantization of codes. (**a**) Distances. (**b**) Indices. (**c**) Quantized code

and 0's elsewhere. As a result, either we look at it from the codebook perspective or the similarity matrix perspective, we should be convinced that indeed we can turn any real-valued vector into a vector consisting of real values but from a finite set. And, most importantly, this whole procedure of quantizing, allows us to apply the backpropagation algorithm! This quantization approach (or very similar) was used in many neural compression methods, e.g., [8, 11] or [10]. We can also use other differential quantization techniques, e.g., vector quantization [16]. However, we prefer to stick to a simple codebook that turns to be pretty effective in practice.

```
class Quantizer(nn.Module):
    def __init__(self, input_dim, codebook_dim, temp=1.e7):
        super(Quantizer, self).__init__()
        #temperature for softmax
        self.temp = temp

        # dimensionality of the inputs and the codebook
        self.input_dim = input_dim
        self.codebook_dim = codebook_dim

        # codebook layer (a codebook)
        # initialize uniformly and a Parameter (learnable)
        self.codebook = nn.Parameter(torch.FloatTensor(1, self.
codebook_dim,).uniform_(-1/self.codebook_dim, 1/self.
codebook_dim))

    # A function for codebook indices (a one-hot representation)
    to values in the codebook.
    def indices2codebook(self, indices_onehot):
        return torch.matmul(indices_onehot, self.codebook.t()).
squeeze()

    # A function to change integers to a one-hot representation.
    def indices_to_onehot(self, inputs_shape, indices):
```

```
21        indices_hard = torch.zeros(inputs_shape[0], inputs_shape
      [1], self.codebook_dim)
22        indices_hard.scatter_(2, indices, 1)
23
24    # The forward function:
25    # − First, distances are calculated between input values and
      codebook values.
26    # − Second, indices (soft − differentiable, hard − non−
      differentiable) between the encoded values and the codebook
      values are calculated.
27    # − Third, the quantizer returns indices and quantized code (
      the output of the encoder).
28    # − Fourth, the decoder maps the quantized code to the
      observable space (i.e., it decodes the code back).
29    def forward(self, inputs):
30        # inputs − a matrix of floats, B x M
31        inputs_shape = inputs.shape
32        # repeat inputs
33        inputs_repeat = inputs.unsqueeze(2).repeat(1, 1, self.
      codebook_dim)
34        # calculate distances between input values and the
      codebook values
35        distances = torch.exp(−torch.sqrt(torch.pow(inputs_repeat
      − self.codebook.unsqueeze(1), 2)))
36
37        # indices (hard, i.e., nondiff)
38        indices = torch.argmax(distances, dim=2).unsqueeze(2)
39        indices_hard = self.indices_to_onehot(inputs_shape=
      inputs_shape, indices=indices)
40
41        # indices (soft, i.e., diff)
42        indices_soft = torch.softmax(self.temp * distances, −1)
43
44        # quantized values: we use soft indices here because it
      allows backpropagation
45        quantized = self.indices2codebook(indices_onehot=
      indices_soft)
46
47        return (indices_soft, indices_hard, quantized)
```

Listing 8.3 A Quantizer class

Adaptive Entropy Coding Model The last piece in the whole puzzle is entropy coding. We rely on entropy coders like Huffman coding or arithmetic coding. Either way, these entropy coders require from us an estimate of the probability distribution over codes, $p(\mathbf{y})$. Once they have it, they can losslessly compress the discrete signal into a bitstream. In general, we can encode discrete symbols to bits separately (e.g., Huffman coding) or encode the whole discrete signal into a bit representation (e.g., arithmetic coding). In compression systems, arithmetic coding is preferable over Huffman coding because it is faster and more accurate (i.e., better compression) [16].

We will not review and explain in detail how arithmetic coding works. We refer to [16] (or any other book on data compression) for details. There are two facts we need to know and remember. First, if we provide probabilities of symbols, then arithmetic coding does not need to make an extra pass through the signal to estimate them. Second, there is an adaptive variant of arithmetic coding that allows modifying probabilities while compressing symbols sequentially (also known as *progressive coding*).

These two remarks are important for us because, as mentioned earlier, we can estimate $p(\mathbf{y})$ using a deep generative model. Once we learn the deep generative model, the arithmetic coding can use it for lossless compression of codes. Moreover, if we use a model that factorizes the distribution, e.g., an autoregressive model, then we can also utilize the idea of progressive coding.

In our example, we use the autoregressive model that takes quantized code and returns the probability of each value in the codebook (i.e., the indices). In other words, the autoregressive model outcomes probabilities over the codebook values. It is worth to mention though that in our implementation we use the term "entropy coding" but we mean an entropy coding model. Moreover, it is worth mentioning that there are specialized distributions for compression purposes, e.g., the scale hyperprior [9], but here we are interested in deep generative modeling for neural compression.

```python
class ARMEntropyCoding(nn.Module):
    def __init__(self, code_dim, codebook_dim, arm_net):
        super(ARMEntropyCoding, self).__init__()
        self.code_dim = code_dim
        self.codebook_dim = codebook_dim
        self.arm_net = arm_net # it takes B x 1 x code_dim and
    outputs B x codebook_dim x code_dim

    def f(self, x):
        h = self.arm_net(x.unsqueeze(1))
        h = h.permute(0, 2, 1)
        p = torch.softmax(h, 2)

        return p

    def sample(self, quantizer=None, B=10):
        x_new = torch.zeros((B, self.code_dim))

        for d in range(self.code_dim):
            p = self.f(x_new)
            indx_d = torch.multinomial(p[:, d, :], num_samples=1)
            codebook_value = quantizer.codebook[0, indx_d].
    squeeze()
            x_new[:, d] = codebook_value

        return x_new
```

```
26    def forward(self, z, x):
27        p = self.f(x)
28        return -torch.sum(z * torch.log(p), 2)
```

Listing 8.4 An adaptive entropy coding model using an ARM class

A Neural Compression System We have discussed all components of a neural compression system and gave some very specific examples of how they could be implemented. There are many other propositions on how these could be formulated, ranging from elaborated neural network architectures for encoders and decoders to various quantization schemes and entropy coding models. Nevertheless, the presented neural compressor should give you a good idea of how neural networks could be utilized for image (or, generally, data) compression.

In Fig. 8.4, we represent a neural compression system where the transformations in Fig. 8.1 are replaced by neural networks together with a (differentiable) quantization procedure. We also highlight that floats are quantized to obtain a discrete signal. Altogether, comparing Figs. 8.1 and 8.4, we can notice that the whole pipeline is the same and the differences lie in how transformations are implemented.

The main difference, perhaps, is that a neural compressor could be trained end-to-end and specialized to given data. In fact, the objective of the neural compressor could be seen as a penalized reconstruction error of auto-encoders. Let us assume we have given training data $\mathcal{D} = \{\mathbf{x}_1, \ldots, \mathbf{x}_N\}$ and the corresponding empirical distribution $p_{data}(\mathbf{x})$. Moreover, we have an encoder network with weights ϕ, $\mathbf{y} = f_{e,\phi}(\mathbf{x})$, a differentiable quantizer with a codebook $\mathbf{c}$, $\hat{\mathbf{y}} = Q(\mathbf{y}; \mathbf{x})$, a decoder network with weights θ, $\hat{\mathbf{x}} = f_{d,\theta}(\hat{\mathbf{y}})$, and the entropy coding model with weights λ, $p_\lambda(\hat{\mathbf{y}})$. We can train the model in the end-to-end-fashion by minimizing the following objective ($\beta > 0$):

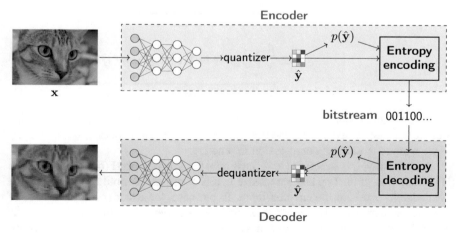

Fig. 8.4 A neural compression system

$$\mathcal{L}(\theta, \phi, \lambda, \mathbf{c}) = \mathbb{E}_{\mathbf{x} \sim p_{data}(\mathbf{x})} \left[(\mathbf{x} - f_{d,\theta}(Q(f_{e,\phi}(\mathbf{x}); \mathbf{c})))^2 \right] +$$

$$+ \beta \mathbb{E}_{\hat{\mathbf{y}} \sim p_{data}(\mathbf{x})} \, \delta(Q(f_{e,\phi}(\mathbf{x});\mathbf{c}) - \hat{\mathbf{y}}) \left[-\ln p_\lambda(\hat{\mathbf{y}}) \right] \tag{8.4}$$

$$= \mathbb{E}_{\mathbf{x} \sim p_{data}(\mathbf{x})} \left[(\mathbf{x} - \hat{\mathbf{x}})^2 \right] + \beta \mathbb{E}_{\hat{\mathbf{y}} \sim p_{data}(\mathbf{x})} \, \delta(Q(f_{e,\phi}(\mathbf{x});\mathbf{c}) - \hat{\mathbf{y}}) \left[-\ln p_\lambda(\hat{\mathbf{y}}) \right]. \tag{8.5}$$

If we look into the objective, we can immediately notice that the first component, $\mathbb{E}_{\mathbf{x} \sim p_{data}(\mathbf{x})} \left[(\mathbf{x} - \hat{\mathbf{x}})^2 \right]$, is the Mean Squared Error (MSE) loss. In other words, it is the reconstruction error. The second part, $\mathbb{E}_{\mathbf{x} \sim p_{data}(\mathbf{x})} \left[-\ln p_\lambda(\hat{\mathbf{y}}) \right]$, is the cross-entropy between $q(\hat{\mathbf{y}}) = p_{data}(\mathbf{x}) \, \delta \left(Q(f_{e,\phi}(\mathbf{x}); \mathbf{c}) - \hat{\mathbf{y}} \right)$ and $p_\lambda(\hat{\mathbf{y}})$, where $\delta(\cdot)$ denotes Dirac's delta. To clearly see that, let us write it down step-by-step:

$$\mathbb{CE} \left[q(\hat{\mathbf{y}}) \| p_\lambda(\hat{\mathbf{y}}) \right] = - \sum_{\hat{\mathbf{y}}} q(\hat{\mathbf{y}}) \ln p_\lambda(\hat{\mathbf{y}}) \tag{8.6}$$

$$= - \sum_{\hat{\mathbf{y}}} p_{data}(\mathbf{x}) \, \delta \left(Q(f_{e,\phi}(\mathbf{x}); \mathbf{c}) - \hat{\mathbf{y}} \right) \ln p_\lambda(\hat{\mathbf{y}}) \tag{8.7}$$

$$= - \frac{1}{N} \sum_{n=1}^{N} \ln p_\lambda \left(Q(f_{e,\phi}(\mathbf{x}_n); \mathbf{c}) \right). \tag{8.8}$$

Eventually, we can write the training objective explicitly, replacing expectations with sums:

$$\mathcal{L}(\theta, \phi, \lambda, \mathbf{c}) = \underbrace{\frac{1}{N} \sum_{n=1}^{N} \left(\mathbf{x}_n - f_{d,\theta} \left(Q \left(f_{e,\phi}(\mathbf{x}_n); \mathbf{c} \right) \right) \right)^2}_{distortion} +$$

$$+ \underbrace{\frac{\beta}{N} \sum_{n=1}^{N} \left[-\ln p_\lambda \left(Q \left(f_{e,\phi}(\mathbf{x}_n); \mathbf{c} \right) \right) \right]}_{rate}. \tag{8.9}$$

In the training objective, we have a sum of distortion and rate. Please note that during training, we do not need to use entropy coding at all. However, it is necessary if we want to use neural compression in practice.

Additionally, it is beneficial to discuss how the **bits per pixel** (bpp) is calculated. The definition of the bpp is the total size in bits of the encoder output divided by the total size in pixels of the encoder input. In our case, the encoder returns a code of size M and each value is mapped to one of K values (let us assume that $K = 2^\kappa$. As a result, we can represent the quantized code using indices, i.e., integers. Since we have K possible integers, we can use κ bits to represent each of them. As a result, the code is described by $\kappa \times M$ bits. In other words, we can use a uniform distribution with probability equal $1/(\kappa \times M)$ that gives the bpp equal $-\log_2(1/(\kappa \times M))/D$. However, we can improve this score by using entropy coding. As a result, we can use the rate value and divide it by the size of the image, D, to obtain the bpp, i.e., $-\log_2 p(\hat{\mathbf{y}})/D$.

```python
class NeuralCompressor(nn.Module):
    def __init__(self, encoder, decoder, entropy_coding,
    quantizer, beta=1., detaching=False):
        super(NeuralCompressor, self).__init__()

        print('Neural Compressor by JT.')

        # we
        self.encoder = encoder
        self.decoder = decoder
        self.entropy_coding = entropy_coding
        self.quantizer = quantizer

        # beta determines how strongly we focus on compression
        against reconstruction quality
        self.beta = beta

        # We can detach inputs to the rate, then we learn rate
        and distortion separately
        self.detaching = detaching

    def forward(self, x, reduction='avg'):
        # encoding
        #-non-quantized values
        z = self.encoder(x)
        #-quantizing
        quantizer_out = self.quantizer(z)

        # decoding
        x_rec = self.decoder(quantizer_out[2])

        # Distortion (e.g., MSE)
        Distortion = torch.mean(torch.pow(x - x_rec, 2), 1)

        # Rate: we use the entropy coding here
        Rate = torch.mean(self.entropy_coding(quantizer_out[0],
        quantizer_out[2]), 1)

        # Objective
        objective = Distortion + self.beta * Rate

        if reduction == 'sum':
            return objective.sum(), Distortion.sum(), Rate.sum()
        else:
            return objective.mean(), Distortion.mean(), Rate.mean
        ()
```

Listing 8.5 A neural compression class

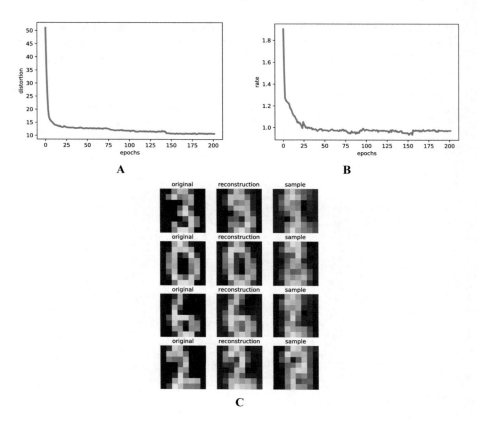

Fig. 8.5 An example of outcomes after the training: (**a**) A distortion curve. (**b**) A rate curve. (**c**) Real images (left columns) and their reconstructions (middle column), and samples from the autoregressive entropy coder (right column)

To conclude the description of the neural compressor, the whole compression procedure consists of the following steps, assuming that the model has been trained already:

1. Encode the input image, $\mathbf{y} = f_{e,\phi}(\mathbf{x})$.
2. Quantize the code, $\hat{\mathbf{y}} = Q(\hat{\mathbf{y}}; \mathbf{c})$.
3. Turn the quantized code $\hat{\mathbf{y}}$ into a bitstream using $p_\lambda(\hat{\mathbf{y}})$ and, e.g., arithmetic encoding.
4. Sent the bits.
5. Decode bits into $\hat{\mathbf{y}}$ using $p_\lambda(\hat{\mathbf{y}})$ and, e.g., arithmetic decoding.
6. Decode $\hat{\mathbf{y}}$, $\hat{\mathbf{x}} = f_{d,\theta}(\hat{\mathbf{y}})$.

Example In a separate file (see https://github.com/jmtomczak/intro_dgm), you can find the implementation of the presented neural compressor and play around with it. For instance, you can get the following results (for $\beta = 1$) as in Fig. 8.5.

Interestingly, since the entropy coder is also a deep generative model, we can sample from it. In Fig. 8.5c, there are four samples presented. They indicate that the model indeed can learn the data distribution of the quantized codes!

8.5   What's Next?

Neural compression is a fascinating field. Utilizing neural networks for compression opens new possibilities for developing new coding schemes. Neural compression achieved comparable or (sometimes) better results than standard codecs on image compression [20], however, there is still room for improvement in video compression or audio compression even though recent attempts are very promising [21, 22]. There are many interesting research directions I missed here. I highly recommend taking a look at a nice overview of neural compression methods [20]. Lastly, I want to highlight that here we used the deep autoregressive generative model; however, we can use other deep generative models (e.g., flows, VAEs).

References

1. Facebook. Facebook reports fourth quarter and full year 2020 results.
2. Rashid Ansari, Christine Guillemot, and Nasir Memon. Jpeg and jpeg2000. In *The Essential Guide to Image Processing*, pages 421–461. Elsevier, 2009.
3. Zixiang Xiong and Kannan Ramchandran. Wavelet image compression. In *The Essential Guide to Image Processing*, pages 463–493. Elsevier, 2009.
4. Yann LeCun, Yoshua Bengio, and Geoffrey Hinton. Deep learning. *nature*, 521(7553):436–444, 2015.
5. Jürgen Schmidhuber. Deep learning in neural networks: An overview. *Neural networks*, 61:85–117, 2015.
6. Lionel Gueguen, Alex Sergeev, Ben Kadlec, Rosanne Liu, and Jason Yosinski. Faster neural networks straight from jpeg. *Advances in Neural Information Processing Systems*, 31:3933–3944, 2018.
7. Lucas Theis, Wenzhe Shi, Andrew Cunningham, and Ferenc Huszár. Lossy image compression with compressive autoencoders. *arXiv preprint arXiv:1703.00395*, 2017.
8. Eirikur Agustsson, Fabian Mentzer, Michael Tschannen, Lukas Cavigelli, Radu Timofte, Luca Benini, and Luc Van Gool. Soft-to-hard vector quantization for end-to-end learning compressible representations. In *Proceedings of the 31st International Conference on Neural Information Processing Systems*, pages 1141–1151, 2017.
9. Johannes Ballé, David Minnen, Saurabh Singh, Sung Jin Hwang, and Nick Johnston. Variational image compression with a scale hyperprior. In *International Conference on Learning Representations*, 2018.
10. Amirhossein Habibian, Ties van Rozendaal, Jakub M Tomczak, and Taco S Cohen. Video compression with rate-distortion autoencoders. In *Proceedings of the IEEE/CVF International Conference on Computer Vision*, pages 7033–7042, 2019.
11. Fabian Mentzer, Eirikur Agustsson, Michael Tschannen, Radu Timofte, and Luc Van Gool. Conditional probability models for deep image compression. In *Proceedings of the IEEE Conference on Computer Vision and Pattern Recognition*, pages 4394–4402, 2018.

12. David Minnen, Johannes Ballé, and George Toderici. Joint autoregressive and hierarchical priors for learned image compression. *arXiv preprint arXiv:1809.02736*, 2018.
13. Claude Elwood Shannon. A mathematical theory of communication. *The Bell system technical journal*, 27(3):379–423, 1948.
14. Yibo Yang, Robert Bamler, and Stephan Mandt. Improving inference for neural image compression. *Advances in Neural Information Processing Systems*, 33, 2020.
15. Fabian Mentzer, George D Toderici, Michael Tschannen, and Eirikur Agustsson. High-fidelity generative image compression. *Advances in Neural Information Processing Systems*, 33, 2020.
16. David Salomon. *Data compression: the complete reference*. Springer Science & Business Media, 2004.
17. LJ Karam. Lossless image compression. In Al Bovik, editor, *The Essential Guide to Image Processing*. Elsevier Academic Press, 2009.
18. Pierre Moulin. Multiscale image decompositions and wavelets. In *The essential guide to image processing*, pages 123–142. Elsevier, 2009.
19. Zhou Wang, Eero P Simoncelli, and Alan C Bovik. Multiscale structural similarity for image quality assessment. In *The Thirty-Seventh Asilomar Conference on Signals, Systems & Computers, 2003*, volume 2, pages 1398–1402. IEEE, 2003.
20. Johannes Ballé, Philip A Chou, David Minnen, Saurabh Singh, Nick Johnston, Eirikur Agustsson, Sung Jin Hwang, and George Toderici. Nonlinear transform coding. *IEEE Journal of Selected Topics in Signal Processing*, 15(2):339–353, 2020.
21. Adam Golinski, Reza Pourreza, Yang Yang, Guillaume Sautiere, and Taco S Cohen. Feedback recurrent autoencoder for video compression. In *Proceedings of the Asian Conference on Computer Vision*, 2020.
22. Yang Yang, Guillaume Sautière, J Jon Ryu, and Taco S Cohen. Feedback recurrent autoencoder. In *ICASSP 2020-2020 IEEE International Conference on Acoustics, Speech and Signal Processing (ICASSP)*, pages 3347–3351. IEEE, 2020.

Appendix A
Useful Facts from Algebra and Calculus

A.1 Norms & Inner Products

Norm Definition

Norm is a function $\| \cdot \| : \mathbb{X} \to \mathbb{R}_+$ with the following properties:

1. $\|\mathbf{x}\| = 0 \Leftrightarrow \mathbf{x} = \mathbf{0}$,
2. $\|\alpha \mathbf{x}\| = |\alpha| \|\mathbf{x}\|$, where $\alpha \in \mathbb{R}$,
3. $\|\mathbf{x} + \mathbf{y}\| \leq \|\mathbf{x}\| + \|\mathbf{y}\|$.

Inner Product Definition

The inner product is a function $\langle \cdot, \cdot \rangle : \mathbb{X} \times \mathbb{X} \to \mathbb{R}$ with the following properties:

1. $\langle \mathbf{x}, \mathbf{x} \rangle \geq 0$,
2. $\langle \mathbf{x}, \mathbf{y} \rangle = \langle \mathbf{y}, \mathbf{x} \rangle$,
3. $\langle \alpha \mathbf{x}, \mathbf{y} \rangle = \alpha \langle \mathbf{x}, \mathbf{y} \rangle$,
4. $\langle \mathbf{x} + \mathbf{z}, \mathbf{y} \rangle = \langle \mathbf{x}, \mathbf{y} \rangle + \langle \mathbf{z}, \mathbf{y} \rangle$.

Chosen Properties of Norm and Inner Product

- $\langle \mathbf{x}, \mathbf{x} \rangle = \|\mathbf{x}\|^2$,
- $|\langle \mathbf{x}, \mathbf{y} \rangle| \leq \|\mathbf{x}\| \|\mathbf{y}\|$ (for a vector in $\mathbb{R}^D$ $\langle \mathbf{x}, \mathbf{y} \rangle = \|\mathbf{x}\| \|\mathbf{y}\| \cos \theta$),
- $\|\mathbf{x} + \mathbf{y}\|^2 = \|\mathbf{x}\|^2 + 2\langle \mathbf{x}, \mathbf{y} \rangle + \|\mathbf{y}\|^2$.
- $\|\mathbf{x} - \mathbf{y}\|^2 = \|\mathbf{x}\|^2 - 2\langle \mathbf{x}, \mathbf{y} \rangle + \|\mathbf{y}\|^2$.

A.2 Matrix Calculus

Liner Dependency

Let ϕ_m be a non-linear transformation, and $\mathbf{x} \in \mathbf{R}^M$. A linear product of these two vectors is:

J. M. Tomczak, *Deep Generative Modeling*,
https://doi.org/10.1007/978-3-030-93158-2

$$\boldsymbol{\phi}(\mathbf{x})^{\mathrm{T}}\mathbf{w} = w_0\phi_0(\mathbf{x}) + w_1\phi_1(\mathbf{x}) + \ldots + w_{M-1}\phi_{M-1}(\mathbf{x})$$

$$= \sum_{m=0}^{M-1} w_m\phi_m(\mathbf{x}),$$

where $\mathbf{w} = (w_0 \ w_1 \ldots w_{M-1})^{\mathrm{T}}$, $\boldsymbol{\phi}(\mathbf{x}) = (\phi_0(\mathbf{x}) \ \phi_1(\mathbf{x}) \ldots \phi_{M-1}(\mathbf{x}))^{\mathrm{T}}$.

Orthogonal and Orthonormal Vectors

Vectors $\mathbf{x}$ and $\mathbf{y}$ are orthogonal vectors if $\langle \mathbf{x}, \mathbf{y} \rangle = 0$. Additionally, if $\|\mathbf{x}\| = \|\mathbf{y}\| = 1$, these vectors are called orthonormal.

Chosen Properties of Matrix Calculus

- $(\mathbf{AB})^{-1} = \mathbf{B}^{-1}\mathbf{A}^{-1}$
- $(\mathbf{AB})^{\mathrm{T}} = \mathbf{B}^{\mathrm{T}}\mathbf{A}^{\mathrm{T}}$
- Matrix $\mathbf{A}$ is positive definite $\Leftrightarrow$ for all vectors such that $\mathbf{x} \neq 0$ the following inequality holds true $\mathbf{x}^{\mathrm{T}}\mathbf{Ax} > 0$
- $\nabla_{\mathbf{x}}\mathbf{x}^{\mathrm{T}}\mathbf{x} = 2\mathbf{x}$
- $\nabla_{\mathbf{x}}\|\mathbf{W}^{\frac{1}{2}}(\mathbf{b} - \mathbf{Ax})\|_2^2 = -2\mathbf{A}^{\mathrm{T}}\mathbf{W}(\mathbf{b} - \mathbf{Ax})$, where $\mathbf{W}$ is a symmetric matrix

For given vectors $\mathbf{x}$, $\mathbf{y}$ and a matrix $\mathbf{A}$, which is symmetric and positive definite, one gets

- $\dfrac{\partial}{\partial\mathbf{y}}(\mathbf{x} - \mathbf{y})^{\mathrm{T}}\mathbf{A}(\mathbf{x} - \mathbf{y}) = -2\mathbf{A}(\mathbf{x} - \mathbf{y})$

- $\dfrac{\partial(\mathbf{x} - \mathbf{y})^{\mathrm{T}}\mathbf{A}^{-1}(\mathbf{x} - \mathbf{y})}{\partial\mathbf{A}} = -\mathbf{A}^{-1}(\mathbf{x} - \mathbf{y})(\mathbf{x} - \mathbf{y})^{\mathrm{T}}\mathbf{A}^{-1}$

- $\dfrac{\partial \ln \det(\mathbf{A})}{\partial\mathbf{A}} = \mathbf{A}^{-1}$

Special Cases of Invertible Matrices

$$\mathbf{A}^{-1} = \begin{bmatrix} a & b \\ c & d \end{bmatrix}^{-1} = \frac{1}{ad - bc}\begin{bmatrix} d & -b \\ -c & a \end{bmatrix}$$

$$\mathbf{A}^{-1} = \begin{bmatrix} a & b & c \\ d & e & f \\ g & h & k \end{bmatrix}^{-1} = \frac{1}{\det \mathbf{A}}\begin{bmatrix} (ek - fh) & (ch - bk) & (bf - ce) \\ (fg - dk) & (ak - cg) & (cd - af) \\ (dh - eg) & (bg - ah) & (ae - bd) \end{bmatrix}$$

Appendix B
Useful Facts from Probability Theory and Statistics

B.1 Commonly Used Probability Distributions

Bernoulli Distribution

$B(x|\theta) = \theta^x (1-\theta)^{1-x}$, where $x \in \{0, 1\}$ i $\theta \in [0, 1]$

$\mathbb{E}[x] = \theta$

$\text{Var}[x] = \theta(1-\theta)$

Categorical (Multinoulli) Distribution

$M(\mathbf{x}|\boldsymbol{\theta}) = \prod_{d=1}^{D} \theta_d^{x_d}$, where $x_d \in \{0, 1\}$ i $\theta_d \in [0, 1]$ for all $d = 1, 2, \ldots, D$,

$\sum_{d=1}^{D} \theta_d = 1$

$\mathbb{E}[x_d] = \theta_d$

$\text{Var}[x_d] = \theta_d(1-\theta_d)$

Normal Distribution

$\mathcal{N}(x|\mu, \sigma^2) = \dfrac{1}{\sqrt{2\pi}\,\sigma} \exp\left\{ -\dfrac{(x-\mu)^2}{2\sigma^2} \right\}$

$\mathbb{E}[x] = \mu$

$\text{Var}[x] = \sigma^2$

Multivariate Normal Distribution

$$\mathcal{N}(\mathbf{x}|\boldsymbol{\mu}, \boldsymbol{\Sigma}) = \frac{1}{(2\pi)^{D/2}} \frac{1}{|\boldsymbol{\Sigma}|^{1/2}} \exp\left\{ -\frac{1}{2}(\mathbf{x}-\boldsymbol{\mu})^T \boldsymbol{\Sigma}^{-1}(\mathbf{x}-\boldsymbol{\mu}) \right\},$$

where $\mathbf{x}$ is D-dimensional vector, $\boldsymbol{\mu}$ – D-dimensional vector of means, $\boldsymbol{\Sigma}$ – $D \times D$ covariance matrix

J. M. Tomczak, *Deep Generative Modeling*,
https://doi.org/10.1007/978-3-030-93158-2

$\mathbb{E}[\mathbf{x}] = \boldsymbol{\mu}$

$\text{Cov}[\mathbf{x}] = \boldsymbol{\Sigma}$

Beta Distribution

$$\text{Beta}(x|a, b) = \frac{\Gamma(a + b)}{\Gamma(a)\Gamma(b)} x^{a-1}(1 - x)^{b-1},$$

where $x \in [0, 1]$ and $a > 0$ i $b > 0$, $\Gamma(x) = \int_0^\infty t^{x-1} e^{-t} dt$

$\mathbb{E}[x] = \frac{a}{a+b}$

$\text{Var}[x] = \frac{ab}{(a+b)^2(a+b+1)}$

Marginal Distribution

In the continuous case:

$$p(x) = \int p(x, y) dy$$

and in the discrete case:

$$p(x) = \sum_y p(x, y)$$

Conditional Distribution

$$p(y|x) = \frac{p(x, y)}{p(x)}$$

Marginal Distribution and Conditional Distribution for Multivariate Normal Distribution

Assume $\mathbf{x} \sim \mathcal{N}(\mathbf{x}|\boldsymbol{\mu}, \boldsymbol{\Sigma})$, where

$$\mathbf{x} = \begin{bmatrix} \mathbf{x}_a \\ \mathbf{x}_b \end{bmatrix}, \qquad \boldsymbol{\mu} = \begin{bmatrix} \boldsymbol{\mu}_a \\ \boldsymbol{\mu}_b \end{bmatrix}, \qquad \boldsymbol{\Sigma} = \begin{bmatrix} \boldsymbol{\Sigma}_a & \boldsymbol{\Sigma}_c \\ \boldsymbol{\Sigma}_c^T & \boldsymbol{\Sigma}_b \end{bmatrix},$$

then we get the following dependencies:

$p(\mathbf{x}_a) = \mathcal{N}(\mathbf{x}_a|\boldsymbol{\mu}_a, \boldsymbol{\Sigma}_a)$,

$p(\mathbf{x}_a|\mathbf{x}_b) = \mathcal{N}(\mathbf{x}_a|\hat{\boldsymbol{\mu}}_a, \hat{\boldsymbol{\Sigma}}_a)$, where

$\hat{\boldsymbol{\mu}}_a = \boldsymbol{\mu}_a + \boldsymbol{\Sigma}_c \boldsymbol{\Sigma}_b^{-1}(\mathbf{x}_b - \boldsymbol{\mu}_b)$,

$\hat{\boldsymbol{\Sigma}}_a = \boldsymbol{\Sigma}_a - \boldsymbol{\Sigma}_c \boldsymbol{\Sigma}_b^{-1} \boldsymbol{\Sigma}_c^T$.

Sum Rule

$$p(x) = \sum_y p(x, y)$$

Product Rule

$$p(x, y) = p(x|y)p(y)$$
$$= p(y|x)p(x)$$

Bayes' Rule

$$p(y|x) = \frac{p(x|y)p(y)}{p(x)}$$

B.2   Statistics

Maximum Likelihood Estimator

There are given N independent examples of $\mathbf{x}$ from the identical distribution $p(\mathbf{x}|\theta)$, $\mathcal{D} = \{\mathbf{x}_1 \ldots \mathbf{x}_N\}$. The likelihood function is the following function:

$$p(\mathcal{D}|\theta) = \prod_{n=1}^{N} p(\mathbf{x}_n|\theta).$$

The logarithm of the likelihood function $p(\mathcal{D}|\theta)$ is given by the following expression:

$$\log p(\mathcal{D}|\theta) = \sum_{n=1}^{N} \log p(\mathbf{x}_n|\theta).$$

Maximum likelihood estimator of the parameters θ_{ML} minimizes the likelihood function:

$$p(\mathcal{D}|\theta_{ML}) = \max_{\theta} p(\mathcal{D}|\theta).$$

Maximum *A Posteriori* Estimator

There are given N independent examples of $\mathbf{x}$ from the identical distribution $p(\mathbf{x}|\theta)$, $\mathcal{D} = \{\mathbf{x}_1 \ldots \mathbf{x}_N\}$. Maximum *a posteriori* (MAP) estimator of the parameters θ_{MAP}

minimizes the *a posteriori* distribution:

$$p(\theta_{MAP}|\mathcal{D}) = \max_{\theta} p(\theta|\mathcal{D}).$$

Risk in Decision Making

Risk (expected loss) is defined as follows:

$$\mathcal{R}[\overline{y}] = \iint L(y, \overline{y}(\mathbf{x})) \, p(\mathbf{x}, y)\mathrm{d}\mathbf{x}\mathrm{d}y,$$

where $L(\cdot, \cdot)$ is the loss function.

Index

© The Author(s), under exclusive license to Springer Nature Switzerland AG 2022
J. M. Tomczak, *Deep Generative Modeling*,
https://doi.org/10.1007/978-3-030-93158-2